SCIENCE AND WONDERS

Russell Stannard is Professor of Physics at the Open University at Milton Keynes. He has travelled widely in Europe and the USA, researching high energy nuclear physics. In 1968 he received the Templeton UK project award and has recently spent a year in America as Visiting Fellow at the Center of Theological Inquiry, Princeton. His best-selling *Uncle Albert* books, which introduce the elements of modern physics to children, have received widespread acclaim.

Science and Wonders

RUSSELL STANNARD

faber and faber
LONDON · BOSTON

First published in 1996
by Faber and Faber Limited
3 Queen Square WC1N 3AU

Printed in England by Clays Ltd, St Ives plc

© Russell Stannard, 1996

Russell Stannard is hereby identified as author
of this work in accordance with Section 77
of the Copyright, Designs and Patents Act 1988

A CIP record for this book is available
from the British Library

ISBN 0–571–17694–1

2 4 6 8 10 9 7 5 3 1

Contents

Acknowledgements

My thanks to producer Norman Winter, who accompanied me on my travels as I gathered material for this book and for the accompanying series on BBC Radio 4. Many of the most provocative and revealing questions put to the interviewees were suggested by Norman.

I owe an enormous debt, of course, to the many people who so generously gave of their time to share their insights with me.

And as always, I would like to thank my wife Maggi for her patience and support.

Those in Conversation

PETER ATKINS: physical chemist and fellow of Lincoln College, Oxford University

IAN BARBOUR: theologian and former physicist, Carleton College, Minnesota, USA

MONTAGUE BARKER: psychiatrist and Medical Director of Heath House Priory Hospital, Bristol

JOCELYN BELL-BURNELL: astronomer, Open University, Milton Keynes, and Presiding Clerk of the Society of Friends

SAM BERRY: geneticist, University College London, and President of Christians in Science

MARGARET BODEN: philosopher and psychologist, Sussex University

SIR HERMANN BONDI: mathematical physicist and President of the British Humanist Association

LAURENCE BROWN: psychologist and Director of the Religious Experience Research Centre, Westminster College, Oxford University

NEIL CHALMERS: biologist and Director of the Natural History Museum, London

JOHN CORNWELL: philosopher and Director of the Science and Human Dimension Project, Jesus College, Cambridge University

PAUL DAVIES: physicist, University of Adelaide, Australia

ROD DAVIES: astronomer and Director of the Nuffield Radio Astronomy Laboratories, Jodrell Bank

RICHARD DAWKINS: zoologist, New College, Oxford

JOHN DURANT: professor of the public understanding of science and Assistant Director, Science Museum, London

JOHN HABGOOD: theologian and former physiologist; former Archbishop of York

PHILIP HEFNER: Director of the Institute for Science and Religion, Chicago, USA

SIR JOHN HOUGHTON: former Director of the Meteorological Office, London

CHRIS ISHAM: theoretical physicist, Imperial College, London

MALCOLM JEEVES: psychologist, St Andrew's University

ERNEST LUCAS: biblical scholar and former biochemist, Baptist College, Bristol

DAVID LIVINGSTONE: geographer, Queen's University, Belfast

NANCEY MURPHY: theologian, Fuller Theological Seminary, Pasadena, USA

RON NAYLOR: philosopher, Greenwich University, London

DAVID PAILIN: philosopher of religion, University of Manchester

ABRAHAM PAIS: physicist, Rockefeller University, New York, USA

ARTHUR PEACOCKE: theologian and former biochemist, Exeter College, Oxford University

TED PETERS: theologian, Lutheran Seminary and Graduate Theological Union, Berkeley, USA

MICHAEL POOLE: visiting research fellow in science education, King's College, London

WILL PROVINE: biologist, Cornell University, Ithaca, NY, USA

STEVEN ROSE: biologist, Open University, Milton Keynes

DAVID ROSEVEARE: chemist, Portsmouth University, and Chairman of the Creation Science Movement in the UK

BOB RUSSELL: theologian and physicist, Director of the Center for Theology and the Natural Sciences, Berkeley, USA

LEON SCHLAMM: psychologist, University of Kent, Canterbury

ANTHONY STORR: retired psychiatrist and freelance writer

ROGER TRIGG: philosopher, University of Warwick

KEITH WARD: theologian, Christ Church, Oxford University

FRASER WATTS: psychologist, Starbridge Lecturer in Theology and Natural Science, Cambridge University

Because spoken conversation does not always read well in written form, it has been necessary on occasion to subject the transcripts of the conversations to minor stylistic editing.

Prologue

Vox pop

The exhibition gallery of the Jodrell Bank Observatory. A school party of youngsters is gathered around illuminated panels displaying photographs of stars and galaxies. They are busily making notes. Having had a word with their teacher, I got chatting with them. I asked them whether they knew how the world came into existence. They shook their heads.

'When you look at all those stars, and at those trees and the grass and everything out there,' I continued, looking towards the window, 'do you reckon someone must have made them?'

'Oh yes,' they readily agreed.

'Who?'

'God!' they chorused, clearly surprised at my not knowing *that*.

'Er ... probably,' a boy murmured hesitantly.

I turned to him. 'Probably? You're not sure?'

He shrugged and looked embarrassed.

'Tell you what,' I said. 'Would you bet your pocket-money on there being a God?'

He brightened up. 'Oh yes,' he agreed. 'I'd do that.'

Those were the views of a bunch of nine-year-olds. They were in marked contrast to what I had encountered a few weeks earlier on a visit to the Natural History Museum in London. There in the main entrance hall, next to the towering skeleton of a dinosaur, I had tackled some older students. First, there were two university undergraduates. They were studying palaeobiology, and had just been upstairs to see the exhibit on evolution. I asked them whether they thought the science they were learning made it easier or more difficult to accept a religious outlook on life.

'More difficult,' said one. 'Science shows the Earth evolved over time, not in six days as religion supports.'

His companion agreed. 'If you want to believe in the Bible, you can. But I mean, you can't disprove what's being shown in here. You can't believe it's just Adam and Eve that started everything.'

I joined a group by the entrance. They were waiting for another member of their party to turn up before leaving. I asked them what they thought of the Bible – having visited a place like this.

'The Bible's simply folk stories, isn't it?' said a young man. 'People believe them if they want to believe them. Personally I don't – any of it.'

A girl nodded. 'The Bible's like fairy stories really. When you're a child, you believe fairy stories, but you don't when you're an adult, do you?'

How did they react to the idea of miracles, I wondered.

'Is it a miracle, or just some scientific fact we haven't yet discovered?' a young woman asked.

'Miracles? It can be coincidence, can't it. A lot of luck. Trickery and that sort of thing,' added her companion.

Overhearing this conversation, others who had been making their way to the exit stopped and were soon joining in:

'You either believe science started it or God started it. If you believe in one, you can't believe the other.'

'I think people believe what they want to believe. There's a lot of illusion in it.'

'I wasn't brought up to believe in God. I'm very much for science.'

'The fossil record shows there can't be any religious explanation of creation.'

'I think that scientists will find out everything as long as they keep on researching and discovering. They'll find the answers to everything.'

'Being a scientist, you *can't* believe the Bible.'

A journey

These exchanges vividly confirmed what has been well known since the 1960s. Surveys have demonstrated that up to the age of eleven, almost all children believe in God and are well-disposed towards religion. But from that time on, attitudes harden. By the late teens, scepticism has taken a firm hold.

Why? There seem to be two contributing factors. The first is that youngsters find it increasingly difficult to understand how a God who is supposed to be all-powerful and loving can allow evil and suffering in the world.

The second factor is science. Young people are learning about evolution in their biology lessons, and they get to hear of Big Bang cosmology. None of this appears to have the remotest connection with what they have been taught from the Bible. So does that not mean the Bible has been caught out? They learn how the workings of nature are not as haphazard as was once thought, but are predictable; they are subject to laws – this makes the acceptance of miracles that much more difficult. The findings of science can be checked out experimentally, leading to universal agreement; this contrasts with religion, where there appears to be greater scope for personal opinion – as witness the proliferation of different religions, and of sects within religions. Science seems always, relentlessly to be advancing; more and more phenomena succumb to its explanations. Might it not one day be able to explain *everything*?

These views, formed in the teens, are then liable to be carried over into adulthood, leading to a widespread belief that science and religion are in conflict with each other. It is commonly held that religion is something you grow out of; that when it comes to the big questions of life – those to do with meaning and purpose, and where we humans fit into the overall picture – we should be looking not to religion, as was done in the past, but to science. Though people still retain a sense of awe and wonder when confronted by the mysteries of the universe, the sense of worship to which this sometimes leads has been displaced from God to science.

But is there really a divide? Is conflict between science and religion inevitable? That's what this book is about. It's a journey. One that took me all over the country, and to the USA. I went to find out what professional scientists themselves thought about the issues. For instance, what effect does a lifetime devoted to using a telescope have on an astronomer? Faced every night by the sheer vastness of the heavens and the multitudes of other worlds out there, does she end up thinking of herself as being entirely insignificant – a meaningless by-product of cosmic events? Does an evolutionary biologist regard himself as no different from the other animals? Does a psychologist believe that all the contents of his own mind, the way he thinks and what he believes, are open to being explained away one day? Does a neuroscientist believe that dissection of the physical brain could one day lay bare everything that he regards as constituting his 'personhood'? Does an expert working on artificial intelligence consider that she is herself nothing more than a kind of robot reacting mechanically to the environment and to instructions programmed into her through the genes she inherited at conception?

And then how about theologians? Do they take an interest in science, and if so, what conclusions do they draw? Are they having to rethink their understanding of God in the light of new scientific developments? And not to be left out, of course, are the philosophers, whose job is to reflect on the nature of reality and to clarify the methods, aims, and limitations of fields of study such as the sciences and theology.

That, then, is the aim: to consider the great questions of life – our origins, purpose, and possible destiny – and to do this in the light of a whole range of expert viewpoints drawn from diverse fields and varying backgrounds.

From the outset, let me declare my own interest: I'm a Christian as well as a scientist – a physicist. Throughout I will try to make clear where I stand personally on the issues (so you can make whatever allowance you think appropriate for possible bias). But as you will see, quite a number of the people I chose to speak with on my journey were those with whom I don't agree: atheists, agnostics – and indeed, certain of my fellow religious believers.

Chapter 1

The Cosmos

The Galileo affair

It is customary to see the supposed parting of the ways between science and religion as dating from the time of the Galileo affair in 1616.

Galileo supported the views of Copernicus: the Earth went round the Sun, not vice versa. He wrote a book contrasting the two viewpoints, *Dialogues Concerning the Two Chief World Systems*. For this, he was brought to trial by the Church. He was made to recant, to disown his scientific discoveries. According to the Church, the Earth was still to be thought of as being at the centre of the universe.

The desire to think of ourselves as being at the centre of things still lives on today, as I found at the first port of call on my journey: the Old Royal Observatory in Greenwich. In the visitors' entrance, there is a notice:

> The position of everywhere else in the world is defined by how far it lies to the East or West of Greenwich. Greenwich Mean Time is the accepted measure for all time on Earth. This small building is the centre of time and space ... the centre of the world.

Standing exactly on the meridian (these days illuminated by a row of red lights), I spoke with someone who has written on the subject of Galileo: Ron Naylor, from the nearby University of Greenwich. How did he account for the affair?

'The long and short of it was that when Galileo wrote this book, he was told he had to present a range of arguments, not just the Copernican one. He had been told he could not present it as a

1

clear-cut case for Copernicanism. The book passed the censors. In fact, two lots of censors – one in Rome and one in Florence.

'Galileo was a brilliant writer: he was a very witty man, and had a wonderful command of the language. And the interesting point there, of course, is that he wrote in Italian. His book was in the vernacular, so it had a much wider potential audience. At the time most of the work appearing was in Latin, so he was breaking with tradition, which maybe was a little dangerous in itself.

'He pointed out that not only were there the good arguments that Copernicus himself had published earlier, but also the very sound arguments that came from the use of the telescope – for example, that the Moon, instead of being a perfect sphere, had mountains on its surface. The telescope also revealed that Jupiter had moons. Galileo pointed out that Venus had phases, just as the Moon has phases; it revolved, and *had* to revolve around the Sun – there was no other way, and people had to admit that.'

'So,' I asked, 'if the evidence was so good, and if the book had passed the church's own censors, why was he brought to trial?'

'There are a number of explanations one can advance. First of all, he *leant* on the censors. He had very good relationships with the main censor in Rome, Riccardi, who was a very nice man. Galileo was continually negotiating what he could, and could not, put in; he wore them down by attrition! He did it by stealth. He got through much more in favour of Copernicus than he was expected to. He was over-enthusiastic about the Copernican theory.'

'Am I right,' I added, 'that Galileo had included an argument the Pope himself had devised – something along the lines that God, being all-powerful, could have arranged all the experimental evidence to look as though Copernicus was right, without him actually being right? Didn't the Pope actually ask to have this included in the book? And Galileo agreed to do this, but then ridiculed it?'

'Absolutely. In fact, it wasn't very sensible, and nor was it well-handled. The Jesuits presented it to the Pope and said "Look, the man's making a fool of you!"'

'So would you say that was really the cardinal reason why

Galileo was put on trial: he'd made fun of the Pope? It was a personal thing?'

'I wouldn't put it that simply. I think it was the aspect of the affair that tipped the scales against him. I think Galileo had pushed his luck: he had taken advantage of a personal relationship he had with the Pope, and one can understand the Pope feeling somewhat let down. Galileo had abused his trust. Galileo misjudged it. The Pope was an extremely clever man, and like many brilliant men of the Renaissance, he was rather vain.'

'And presumably the Pope was concerned that his *office* was being made fun of?'

'He couldn't have *not* thought that.'

'So would you say that if Galileo had handled the situation more prudently, we needn't have had this trial and all its later ramifications?'

'I think that's quite possible.'

So much for the past. What did Naylor think was the legacy of the Galileo affair?

'I think it's had a lasting legacy and a significant one. I mean, people *do* think that the Church is inevitably apprehensive about change: it doesn't react favourably (that's the impression) to radical changes in science and in the understanding of the Universe – it tends to be conservative.

'The Church obviously learned a lot from the Galileo affair: it didn't come out openly against Darwin! The Catholic Church reserved its position on *that*. But nevertheless, it's left an impression, wherever European cultures have gone, that there is a natural – an inevitable – tension between an authoritarian Church and an activity like science where new ideas have to be openly discussed. There shouldn't be any attempt to constrain discussion, and the Church definitely did attempt to do this.'

One feature of the Galileo affair which has always intrigued me is the fact that the cosmology being defended by the Church at the time was *not even* the view put forward by the Bible. In the Bible we have a three-tiered Universe, with a flat Earth sandwiched between Heaven above and Hell below. But what the Church was defending was a cosmology based on *spheres*, with the Earth at the centre. It was a picture taken over from the Greek philosophers.

So the question is: How come there was all that fuss over Galileo when there had already been a previous drastic change in cosmological thinking, one that had taken place without any bother at all? I can't help thinking the Galileo affair was totally unnecessary. It was a matter of politics and hurt personal feelings, rather than a fundamental rift between science and religion. It would never have happened had it been handled differently, on both sides.

The scale of the Universe

The adoption of the Copernican viewpoint was but one step towards a more adequate conception of the Earth in relation to the rest of the cosmos. What is our current view of the Universe?

Just a couple of doors down the corridor from my office at the Open University is Jocelyn Bell-Burnell. She is an astronomer, well known for her momentous discovery of 'pulsars'. These are the condensed remnants of old, burnt-out stars. They spin like a top and emit a beam of radiation. Each time the beam sweeps round in our direction we pick it up as a regular pulse. It's rather like a lighthouse beam being swung round. Jocelyn has the distinction of being the only woman professor of physics in the country at the present time. So how would she describe the Universe?

'It's very big and consists of a very large number of things,' she began. 'Our Sun is a star, and we now know that it's a very ordinary star, as average as they come. It is one of something like 100,000 million stars in our galaxy, which is called the Milky Way. And there are a lot of other galaxies like ours. There's a very handy multiplication table we owe to Sir Arthur Eddington: 100,000 million stars equals one galaxy; 100,000 million galaxies equals one Universe.

'The Sun is about 100 million miles away from us. The next nearest star is about two million million miles away from us. Our galaxy is something like 50,000 million million miles across. And some of the most distant things we can see are thousands of millions of millions of millions of miles away. You can't envisage it; it's too big.'

'With all these stars being "suns", do we envisage that quite a number of them have planets?' I asked.

'There's not a lot of hard data to go on, but there are so many stars like our Sun that, even if only one in a hundred has planets, it means an awful lot of planets. And even if only one in a hundred of those planets is at a suitable distance from its star to support life, that's still a lot of stars with planets able to support life.'

The American cosmologist Steve Weinberg, in his introductory book on cosmology *The First Three Minutes*, ended up by saying: 'The more the Universe is comprehensible, the more it seems pointless.' He also described life as 'a more-or-less farcical outcome of a chain of accidents'. I reminded Jocelyn of this and asked: 'When you're faced with the enormity of the Universe, do you personally think that life is important?'

'On a local scale, yes,' she replied. 'I think it is extremely important. But on the cosmic scale, to be honest, I'm not quite sure what it all means.'

She must have noticed me looking puzzled, for she added with a smile, 'That wasn't the answer you were expecting, was it?'

Knowing her to be one of the central figures in the Quaker movement, I wasn't expecting quite such a guarded reply. I continued: 'We read in the Bible how the psalmist looked up at the night sky and saw it as displaying the splendour of God. Today, perhaps, we tend to look up at it with our modern understanding of just how big it all is, and we see it as a reflection of the smallness of human beings. The emphasis seems to have changed, don't you think?'

'There are people who will say "Gosh, isn't that beautiful! It proves there's a God, doesn't it?" I don't like that argument.'

'I don't like it either,' I agreed. 'But I would say that, if you *already* believe in God – on other grounds – then you can look up at that sky and learn something about that God: his enormous power and splendour. Wouldn't you say that?'

'It depends on what you believe about a creator God, doesn't it?'
'True.'
'I think maybe we differ on that as well.'
'You don't see God as the creator of the Universe?' I asked.
'Again, I'm not sure. I wouldn't want to say too categorically that God created it. It could have just *happened* – which is not to say there isn't a God. I do believe there is a God.'

The scientist who has most influenced our understanding of the Universe is, of course, Albert Einstein. He often referred to God. For example, in speaking of quantum theory, he expressed the view that 'God does not play dice'. But what did he actually mean when he used the word 'God'? Abraham Pais of Rockefeller University, New York, himself a physicist, has written extensively about Einstein's life and work. The title of one of his biographies, *Subtle is the Lord*, is itself taken from a quotation of Einstein's.

I asked Dr Pais whether Einstein actually did believe in God, a personal God – or was it just a colourful way of referring to the laws of nature?

'Let me go back to his youth,' he replied. 'When Einstein was about six years old he went through a period of intense religious feelings – conventional religious feelings. He did not eat pork, things like that. That lasted only for a year, and after that it was all gone – he did not believe in a personal God, a God that punishes or rewards.'

'So what did he mean when he used the word "God" in later life?'

'Oh, an almost abstract concept. He meant that nature manifests itself in ways of great regularity; there are laws of nature. And these laws have to be discovered and uncovered. Einstein meant it in that sense – nothing more and nothing less. Apart from that short phase in his childhood, he was never a religious man in his whole life.'

I decided next to call on Peter Atkins. A physical chemist, he is someone who has written extensively about the creation of the Universe from an atheistic point of view. Peter and I have, on occasion, had cause to clash in open debate over the subject of science and religion. His study overlooks the quadrangle of Lincoln College, Oxford. I wanted to know how he reacted to astronomy. Did the size of the universe make him feel insignificant?

'I've always thought that I was insignificant. Getting to know the size of the Universe, I see just how insignificant I really am! And I think the rest of the human race ought to realize just how insignificant it is. I mean, we're just a bit of slime on a planet belonging to one sun. There could be billions of them. There could be other universes. It makes us really, totally insignificant – in one sense.

'But we are possibly highly significant in another. We have the precious ability to observe and to wonder about the Universe. It's conceivable that we are alone, in which case we have a very special responsibility to keep this flame of understanding alive. If we are not alone, then in a sense we don't matter at all; it's entirely irrelevant whether we're here or not. It's just very nice to be here.'

'Do you live your life as though you are insignificant; as though it doesn't really matter what you do, or whether you are here or not?' I asked.

'On a private scale of things, I think it's highly significant that I'm here – for myself at least, and those who are close to me. We have a wonderful opportunity through an accident of birth to enjoy the wonders of the Universe. My day-to-day motivation is to understand more and more about the workings of the world, and to see how it hangs together, what makes it tick. Beyond that, I also take normal human pleasures – pleasure from music, from poetry, and from all the other titillations of the brain that people have devised.'

Next I visited the home of David Pailin, Head of the Department of Philosophy at Manchester University, a specialist in the philosophy of religion. I asked him whether he thought the scale of the Universe had more to say about the greatness of God or the insignificance of ourselves.

'I think it's a bit of both,' he replied.

'If one of the effects of contemplating the cosmos is to make us human beings feel more humble, do you think we nevertheless retain some dignity? If so, in what does that dignity lie?'

'Certainly we have dignity,' he answered firmly. 'The dignity lies in being persons whose personal identity is respected by God, and of whom God is individually aware. This is one of the great and significant things about the sacrament of baptism: someone is given a name *as an individual* and is recognized by God for the person that they are.'

'Is there any scientific advance which you think has been of particular significance for theologians and philosophers of religion?'

'We are becoming aware that our planet is just one very, very small piece of a huge story. The notion of God as creator is really a mind-boggling idea. It needs to be radically rethought. So often we tend to think that all of God's creative activity, and everything else, was put up for human benefit. We have got to think of God as the God of the cosmos. We are still having difficulties in realizing that the human community on this planet is only one small part of that cosmos.'

Ian Barbour was a physicist before becoming a theologian, and now lectures at Carleton College in Minnesota. He is one of the most respected scholars in the field of science and religion. Like myself, he is on the Board of Advisers to the John Templeton Foundation, a charitable organization devoted to promoting progress in religion. It was following one of the Board's meetings, at a hotel in Atlanta, that he explained to me that, when it comes to 'significance' what counted most for him was the notion of complexity:

'We're clearly in a much larger and longer-lived cosmos than we had originally thought. But I'm not sure that we need to conclude that we're insignificant. It is an immense cosmos, but it had to be immense for life and consciousness to evolve. That takes a long time, and if you've got an expanding universe – and a long time (fifteen thousand million years) – it's bound to be a big universe.

'I'm not sure, though, that size is the best measure of importance. It seems that the most complex and interesting kinds of thing occur at a middle scale, neither way down at the very small size of atoms, nor at the enormous size of galaxies. There's more complexity in the human mind than in a thousand lifeless galaxies or planets. Consciousness and complexity are perhaps more important than sheer size. After all, it's *we* who are thinking about all that size, whereas on a lifeless planet that kind of thought isn't going on.'

It occurred to me that this was very close to what Peter Atkins had been saying. He too saw a role for us in keeping 'the flame of understanding alive', as he put it. The fact is that, thanks to the complexity of our brains, we are conscious, whereas an object like the Sun is not. Though there can be no denying the greater importance of the Sun when it comes to physics, one would not swap places with the Sun. That in itself must surely be saying something. What's the point of being important if you don't know it – if, in fact, you don't know anything at all?

The French philosopher Blaise Pascal summed it up well when he said: 'Man is only a reed, the weakest thing in nature; but he is a *thinking* reed.'

The origin of the Universe

As we have seen, the Universe is large. But was it always like that? Ian Barbour mentioned in passing that it was an *expanding* universe. What do we know about the origins of the Universe and how it has subsequently developed? Jocelyn Bell-Burnell again:

'Well, none of us was there to see it, but we think what happened was that there was an enormous Big Bang. Initially, everything was concentrated into a very, very dense spot (for want of a better word) and this exploded. And out of that explosion came everything: all space, all time, all matter, all energy. And since then it has been gradually expanding and cooling. Initially, it was a great ball of energy, but with time that has turned into matter. It all came from the Big Bang some fifteen or twenty thousand million years ago.'

There are essentially three pieces of evidence for the Big Bang.

The first is that the galaxies are still moving apart as a result of that great explosion. We can tell this because the light we get from them is redder than one would otherwise expect. This is what you get if a source of light is receding from you (rather in the same way as the pitch of a police car's siren is lower the faster it moves away from you). So the galaxies are moving apart. But not only that, the fastest-moving galaxies are those now furthest away from us – which is what one would have expected if initially all the material had started off together. In fact, there is a neat relationship between the speed of recession of a galaxy and how far away it is: they are proportional to each other. This relationship involves the famous *Hubble's constant*. From it one can extrapolate back in time to find out when all the matter must have been together, i.e. when the Big Bang occurred.

Secondly, as with any violent explosion, the Big Bang was accompanied by a blinding flash of light. The cooled-down remnant of that flash is still about in the Universe today, and has now been detected. It is called the cosmic microwave background radiation.

Lastly, from an understanding of the kind of nuclear processes that go on in the intensely hot conditions which must have occurred during the very early stages of the Big Bang, it is possible to work out what proportion of the different types of atom we would have expected to have come out of the Big Bang, and thus what one would expect to find in space today. This too checks out with what is found in practice.

So there are three independent pieces of evidence, and they all point to the same conclusion. It is not surprising, therefore, that there is almost universal agreement that the Universe began with a Big Bang.

I say 'almost' because there are those who do *not* accept this conclusion. These include the members of the Creation Science Movement, a UK organization devoted to upholding the literal interpretation of Genesis, with its account of the creation of the world in six days. David Roseveare, himself a scientist (a chemist at Portsmouth University), is chairman of the Movement. I met up with him in a radio studio in Broadcasting House, London, to try to find out how he could possibly hold his beliefs in the face of the evidence.

He began: 'When the Big Bang theory was first postulated, it was rejected because the age it gave was insufficient. Even today, the Hubble constant is revised from time to time. The age of the Earth might be doubled, or halved, at a stroke ...'

He was drawing attention to the fact that there are difficulties in gauging the distance to far-off galaxies. If new evidence points to a need to revise the estimate upwards, say, then that means the galaxy has travelled further than originally thought, and that in turn implies it has been travelling for a longer time – hence the Big Bang must have happened earlier than originally thought. I agreed with him that there had been revisions of Hubble's constant.

'But,' I put it to him, 'we're talking about revisions by factors of two or three, that sort of figure. To talk about comparing a 15,000 million-year life of the Universe with a 6,000-year life (which a literal interpretation of the Bible would indicate) means a factor which is so absolutely enormous that you're never going to be able to explain it away.'

He changed tack.

'There are many other problems with the Big Bang theory,' he asserted. 'The idea of a free meal, whereby once upon a time nothing exploded into everything, goes against all our expectations. You don't get something for nothing. The idea that the orderliness in our Solar System could arise from the explosion of everything is again surprising; we tend to feel that explosions cause chaos rather than order, and yet we all set our clocks by the movement of the heavenly bodies.'

'I agree that no Big Bang theory – in fact, no science at all – can explain how something can come out of nothing. That is going beyond the realm of science. But I would have thought that the Big Bang theory was now extremely well established. You have three independent pieces of information. You've got the recession of the galaxies, which points back to them having all been in one place at one time. You have the remnants of the fireball of the Big Bang. It has been measured, and it has exactly the characteristics you'd expect. We can measure the abundancies of the different kinds of atoms in the Universe; they have exactly the proportions you would expect to come out of the conditions of the Big Bang. I find it extremely strange that we get these three apparently independent pieces of information and they have nothing whatsoever to do with the Genesis account of something happening 6,000 years ago.'

'Let's take these three lines of evidence that you refer to. Firstly, the evidence that the galaxies are moving away from one another. It's been suggested by several people that there are other explanations for the red shift of light from distant galaxies. The idea of the background radiation is also not so foolproof ...'

He carried on for a while asserting that the evidence was not as conclusive as is generally held, but I remained unconvinced. I asked: 'Why do you think God created the world in such a way as to build into it false clues about how the Universe came into being? What kind of God do you believe in? One who apparently set out deliberately to deceive the vast majority of humankind as to how they came about?'

'Well, let's be clear. God lays it down very clearly in Genesis 1 that he did it in six days. If people interpret the evidence in another way, mainly in order to give a naturalistic view which allows us to cut God out, or to use God simply to wind up the mechanism to start with, one can hardly blame God for that.'

'Would you agree that there are ways of conveying truth other than by speaking literally?' I asked. 'Can poetry, for example, convey a truth about ourselves? Doesn't it worry you that perhaps God, apart from all his other attributes, was also a poet, and what we find in Genesis is his poetry – a poetry that has to be interpreted, just like any poem has to be interpreted?'

'There's a great deal of poetry, of course, in the Scriptures. The Book of Psalms, and so on. Genesis, generally speaking, is an historic book. If we take the portion from Genesis 12 through to the end, we find that it's generally regarded as the history of Abraham; it's a history leading through to the early patriarchs of Israel. If we come back to before then, we find the story of Noah, which again has always been regarded as history by the early Church. The Lord

Jesus himself spoke of his coming judgement in such terms as "In the days of Noah ..." And when we come to Genesis 1 and Genesis 2, they are not written in poetic form.

'There is one verse in Genesis 1 (verse 27) which is poetry. It says, "In the image of God created he him, male and female created he them". I'm very pleased that God writes poetically about making man in his own image. And when you come to Genesis 2, there is again just one verse which is poetical in structure. That's where Adam looks at Eve and says, "This is now bone of my bone and flesh of my flesh". Again, I'm pleased that he waxed poetical about seeing his mate. But in general you have an historic form to the literature.'

And so our conversation progressed – or, perhaps more accurately, failed to progress. There was simply no meeting of minds here. Literalist, fundamentalist Christians have always bothered me. On the one hand, they clearly have a deep respect and love for the Bible, which I unreservedly applaud. But theirs is an approach that appears to fly in the face of the scientific evidence. While one must always be on one's guard against making exaggerated claims for scientific *certainty*, there is surely such a thing as a reasonable and balanced judgement based on scientific evidence. Theirs is an approach that brings their religion into headlong collision with such sober scientific judgements. As we shall see in our later discussion of evolution, the creationist movement remains powerful, especially in the USA. Sadly, its activities lead to a significant number of scientists becoming contemptuous of *all* religion.

I stated in my conversation with Dr Roseveare that the question of how something could come out of nothing lay beyond the remit of science. But how close to an explanation can we get? Does it make sense to ask, 'What caused the Big Bang?' Someone well known for his explanations of difficult physics is Paul Davies from Adelaide University, Australia. What did he think?

'If you take a simple-minded picture of this Big Bang, then the originating explosion itself is not a normal physical event,' he agreed. 'When I was a student, you had to accept the origin of the Universe in a Big Bang as simply given. In recent years, a number of people have attempted to bring the very origin itself within the scope of physical science by appealing to quantum physics. This is a weird and wonderful branch of science which applies primarily to atoms and molecules, to little things. The Universe is a big thing, the biggest thing we can imagine. But if we're right that it has expanded from a very small size, then there must have been a time sufficiently early on when quantum effects would have been important.

'Now, the essence of quantum physics is something called Heisenberg's Uncertainty Principle. This, roughly speaking, says that everything that can be observed or measured is subject to unpredictable fluctuations, to uncertainty. In the laboratory, this leads to circumstances where events can occur without well-defined prior causes. That is to say, in the words of Tommy Cooper, they happen "just like that"! We see, for example, the nuclei of radioactive atoms decaying "just like that". If you ask, "Why did it decay at that moment, rather than some other?" there is no answer. It is a genuinely spontaneous event. This is not quite the same as total lawlessness, but it does introduce spontaneity into nature, at the microscopic level, in a very fundamental way.

'Of course, it's a huge extrapolation to say that, given this spontaneity, we can explain how the Universe as a whole can come into being "just like that" – through a spontaneous quantum event. We really have two choices when it comes to the origin of the Universe. One is to say it was a supernatural event, and just leave it at that; we're not going to enquire. The other is to find some way of treating it as a natural event. But by definition, if a natural event can occur once, it can occur many times, perhaps an infinite number of times. I think we are inevitably led to expect that if we want to appeal to quantum physics – or anything else – to explain the origin of the Universe as a natural event, then there will possibly be an infinite number of *other* universes that can originate in this same manner. We're making huge assumptions about the nature of reality in postulating these other universes. But it seems that one is really forced into that type of picture by any attempt to treat the origin of the Universe as a natural event.'

So that is a physical mechanism by which the Big Bang might have come about. It is an explanation (of sorts) that is currently exciting much speculative interest in scientific circles. But at the same time, such talk can be deeply disturbing to religious believers. A physical explanation of the origins of the world? Where does God fit in? Isn't God supposed to have created the world?

Peter Atkins is only too happy to dispense with the notion of a 'creator God': 'I think it's one of the several legs on which belief in God rests. If physics can show that one can account for the creation without invoking a creator of any kind, then that is one of the legs sawn away; it's one more reason why one shouldn't believe in a creator God. One of the extraordinary things that has happened over the last twenty-five years is that scientists have shown that they can get extraordinarily close to the beginning of the Universe. We can be very confident about tracing the history back to within about one-billionth of a second after the Universe began. We can be moderately confident that we can trace it back to a period billions of times shorter than that. At no point do we find that we need the involvement of a superior being of any kind. At each point, we can take the physics that we have discovered in the last two or three hundred years and apply it, in a cautious but fascinating way, to the events that took place when the Universe was extraordinarily young.

'Now there comes a point, of course, where physics currently doesn't work. We can't get right back to the instant of creation. But there is no reason to believe that we won't be able to bridge that tiny remaining gap and really understand how the Universe could come into existence without any intervention – without anyone doing anything to make it come into being. We would then be able to see how it came from absolutely nothing. And if we can do that, then that seems to be a pretty strong argument for saying that you don't need a creator to do it.

'What physics has been doing over the last twenty-five years is to show that the creator could be very lazy. As the years have gone by, we've shown that he could be lazier and lazier. In my view, in due course we shall show that the creator could have been infinitely lazy and not have done anything at all! That's what I think will be the final argument against the existence of a creator God.'

That's all very well, but are we looking in the right place for finding God at work – by concentrating on the very instant of the Big Bang?

Creation stories in the Bible

Keith Ward is a theologian. Although he also has a room overlooking the quadrangle of a college of Oxford University (Christ Church), and so is geographically just around the corner from Peter Atkins, he could hardly be further removed from him in his thinking.

'The Big Bang has not the slightest theological significance,' he asserted. 'It doesn't matter whether the world began with a bang or whether it began at all. It might always have been there. Theologians have always agreed that this doesn't make any difference. When you say that God "creates" the Universe, you just mean that everything – however long it's been there – always depends, at every moment of time, upon God. That's all you mean. It's a pity that some physicists use the word "creation" when they actually mean "origin" of the Universe. Creation isn't at the beginning; it's now and always.'

He went on to explain how this understanding has consequences for the way we are meant to read the creation stories in religious scriptures, such as those of the Jewish, Christian and Islamic traditions.

'The stories in the Bible are setting out spiritual meanings of very great depth and significance. They're not meant to be literally true. And this is not a new theory: it was Augustine's view. They obviously couldn't be literal. A very simple point: there couldn't be any "days" before there was a Sun to have day and night periods. The people who wrote Genesis were not stupid. They knew that for the first few days there wasn't any Sun, so they didn't mean "day" literally. When people say they believe that literally, they probably haven't quite thought it through.'

John Habgood was still Archbishop of York when I visited him at Bishopthorpe, just south of York. The chapel in the palace used to have a lofty ceiling. A false floor has now been installed to convert the upper part into a study for the Archbishop. It has a wonderful open view from which one can watch the barges gliding by on the River Ouse below.

I asked him what, if anything, he saw as the theological significance of our understanding of how the world came into being. He confirmed what Keith Ward had told me: 'Theologians on the whole have been fairly relaxed about this. They've said that various scientific descriptions of the origin of the Universe are compatible with an understanding of God as creator. So when we talk about God as creator we are saying that in the end, the existence of the Universe depends upon God. It's not self-sufficient; it's not self-explanatory. But precisely how or when it came to be is really a matter for scientists to work out.'

'But what about the creation stories we find in Genesis?' I enquired. 'Do you see those as having any further relevance for us?'

'Yes, I think they are profoundly interesting, but not as a quasi-scientific description of what might have happened. They give a sense of the orderliness of creation, its dependence on God. The God the Israelites worshipped was the God of the whole Earth. This they expressed in a marvellous piece of poetry.'

'Am I right,' I asked, 'that this idea of God being the God of the whole Earth was a development from earlier ideas of God – ideas that he was just a tribal God, interested only in a particular part of the world and a particular people?'

'Indeed. The more the Israelites became caught up in the great historical events of the day, so their faith in God expanded to meet those demands, until they thought there must be only one God.'

To find out a little more about how one might view Genesis, I next saw Ernest Lucas. He used to be a scientist before taking up

Biblical Studies. He now lectures on the subject at the Baptist College, Bristol. As we talked, we had in front of us the opening words of the Bible. If these were not meant to be taken as literally true, what were they getting at?

He began by explaining that, although Genesis 1 is the first thing one reads on opening the Bible, it's actually quite a late piece of writing, dating from only about six centuries before the birth of Christ, a time when many of the Jewish people were in exile in Babylon. The important thing to recognize is that it's answering different questions from the ones posed by scientists today.

'What the cosmologists are telling us is how they think the cosmos as we know it came into being,' he said. 'The emphasis is on the "how", what the stages were in the development from the Big Bang to the present day. What the writer of Genesis 1 is grappling with are much more questions of "why". Why is there is a Universe, and what sort of nature does that Universe have?

'In these opening verses we're told, "In the beginning God created the heaven and the Earth. And the Earth was without form and void; and darkness was upon the face of the deep ..." That phrase "without form and void", and then the six days of creation – the first three spent with God forming the world by acts of division (light from darkness, the waters below the heavens from those above the heavens, dry land from water), and then three days of filling the void so it's no longer empty – that was saying something very profound in the context of Babylon.

'The Babylonian creation story starts with waters of chaos and monsters of chaos. The creator god – in their terms the god Marduk (the god of Babylon) – has to subdue chaos. It's been said of the Babylonian religion that it was anxiety-ridden; they were always afraid the chaos would break in again. And we all know that feeling from our own experience of life: there are times when you think "Gosh, everything's falling apart". The Hebrew creation story starts from a fundamentally different point of view. This is an ordered cosmos. There's no battle with chaos. Later on, we're told things do go wrong. But you start with an ordered, reliable cosmos.'

Just like that!

I was still concerned about this idea that quantum physics was in charge of bringing the Universe into existence. It seemed all very well saying that the Big Bang could have happened as a spontaneous quantum event – 'just like that', as Paul Davies put it so graphically. But doesn't that lead on to another, more probing question, namely: why quantum physics? Why was it the laws of quantum physics were in charge, rather than some other kind of physics? How come the Universe is the particular kind of universe it is? Does it take a 'God' to decide?

Jocelyn Bell-Burnell, though believing in God, was a little cagey over whether God created the world. I tackled her on whether she thought her God was responsible for setting up the laws of physics, the quantum laws that might then have run their course and created the Universe.

'I accept that you're pushing the question one step further back,' she said.

'So, given that you're unsure about whether God is the creator, are you less unsure of God being the person responsible for deciding what kind of laws will be in charge?'

'That for me is actually the same question. So when I say that I don't know that God created the Universe, I'm also saying I'm not convinced that God set up the laws of physics and lit the blue touch-paper.'

'Do you have any alternative?' I asked.

'No. I don't actually feel the need to have answers to everything. I'm quite happy to work with "don't knows", and I'd rather work with "don't knows" than unnecessary assumptions. It doesn't matter to me, to the way I live my life, whether God created this or not. What matters is that there is a God.' A pragmatic approach to life that certainly serves her well on a day-to-day basis.

But while Jocelyn was clearly not going to lose any sleep worrying over this issue, I still felt a need to pursue the question. So how did Peter Atkins think these laws originated?

'There are various approaches to this,' he began. 'First of all, it's possible that there are other universes in which the fundamental particles have different characteristics. You might have pink electrons in one, blue electrons in another, electrons the weight of elephants in a third, and so on. It just happens that we are inhabiting this particular universe where electrons are much lighter than elephants.

'Another possibility is that there is only one conceivable universe. When a universe comes into existence, it can come into existence only with the particular characteristics that we experience. The only possible electrons are those that have the characteristics we are familiar with. When the Universe tumbled out of nothing, there was just one sort of electron that could emerge from that instant of creation; it was not possible to have a universe with electrons of a different mass. A universe can exist only if it is self-consistent, in the sense that everything hangs together. It may be that, in order for the universe to survive its birth, only one set of properties could exist.'

That, I suppose, is a possibility. But I'm not sure how Peter could ever hope to demonstrate that there can only be this one type of universe. After all, it seems only too easy to dream up other types of world run on different lines. Science-fiction writers are doing it all the time. What did Paul Davies think?

'I'm convinced the Universe *could* have been otherwise,' he replied. 'There could have been different laws.'

'So what stopped it being otherwise? Was there a "decision" of some kind?'

'The fact that we live in the kind of universe we do is very suggestive that not only has a selection been made, but a rather intelligent selection. There are so many remarkable features of the laws that characterize this universe that it's very suggestive of something like design, or meaning, or purpose.'

Design, meaning, purpose? Is this where we see *life* fitting into the scheme of things?

Life in the Universe

My enquiries into the origins of life began in the Nuffield Laboratories at Jodrell Bank, Cheshire. There I met up with the Director, Rod Davies (no relation to Paul Davies). The view from his office window is totally dominated by the famous radio telescope, a marvel of engineering. It has a collecting dish 250 feet in diameter, yet that great bulk can be delicately steered so as to pick up radio signals from any direction.

'What would you say has been the distinctive role that this laboratory has played in our modern understanding of astronomy?' I asked.

'I'd say, looking back over the fifty years the laboratory has been here, that the major contribution to astronomy has been looking at the most distant objects in the Universe. In fact, they have been found to be at something like 50 per cent of the distance to the edge of the Universe, and even further than that – the most distant ones are probably in the order of 80 per cent of the way to the edge of the Universe.'

'How big do we think the Universe is these days?'

'I can't give a very definitive answer. The figure generally used is that the radius of the Universe is about 15 billion light years.'

'A light year being the distance light travels in a year.'

'Precisely that.'

I asked him how he reacted to having to work on a daily basis with such mind-numbingly large distances.

'Well, I think you're in a position where you are certainly looking up at the Universe as a large scale. But if you look down at the atoms, they are awfully small. So I think we are very privileged where we are!'

I knew him to be a Christian, so I asked, 'Do you see any God-given purpose to life in the light of your studies here?'

'I think that's a personal reaction you might have to seeing the Universe on a large scale, and being a part of the creation process. I think that's a religious experience, to look at the whole creation process and to see yourself as a part of it.'

'Do you have that experience yourself?'

'Yes. That is very much my view of the Universe.'

I asked whether there was any aspect of his observations that could throw light on how life originated.

'Using the telescope here, and others working at higher frequency,' he said, 'some hundred different types of chemical molecule have been found in the spaces between the stars. They are of the form of the building-blocks of life. The building materials for life are distributed right through our own galaxy and many other galaxies.'

So with the raw materials for life abundantly spread out through space, how was it converted into living creatures? Paul Davies said, 'I don't believe that the origin of life was a miracle, nor do I believe it was a stupendously improbable accident. I think it's something which is written into the basic laws of the Universe in a very fundamental way.'

To think of life as somehow fundamentally a part of the Universe, as opposed to being some accidental by-product, is not easy. I put it to him that most of the places in the Universe are either too hot or too cold to sustain life. In view of this, it could be argued that the Universe was a rather hostile kind of place. Also, as we've just heard, it's mind-bogglingly big. If it was primarily meant as a home for life, was it not a case of over-design perhaps?

'It's true that life could not exist in most places in the Universe,' he agreed. 'But a lot of the things which seem rather hostile or terrifying are actually prerequisites for life. For example, life on Earth is sustained by the Sun; it's the source of our energy. But, of course, that very heat which is beneficial to the Earth would be intolerable if you got any closer to it. The size of the Universe, likewise, is a prerequisite for life. Life takes a long time to evolve, and if the Universe was a lot smaller, it would also be a lot younger. So when you start investigating these features of the Universe that look so terrifying to ordinary people, you find that they're not so terrifying after all.

'It is actually rather remarkable to sit down with a list of the laws of physics and the various features of the Universe. If we had the power to take a God's-eye view, and could change these laws and conditions in some way, what we find is that even the slightest changes to the underlying laws, or the way the Universe originated, are likely effectively to wreck it ...'

'As far as having life is concerned?'

'... as far as having life, or indeed any complex structures, are concerned. The existence of something as complex as life seems to be very delicately dependent upon a felicitous combination of the underlying laws and conditions.'

Here indeed we come across one of the most amazing things about the Universe, something that has only come to our attention in recent years: the fact that if you were to sit down and draw up a blueprint for an imaginary universe – one in which you just put together a whole set of made-up laws of physics – your chances of having a universe that was capable of leading to the development of intelligent life-forms would be virtually zero. By 'made-up laws' I mean ones in which you make the force of gravity a bit stronger than it really is, or a bit weaker; or you make the heat-generating nuclear reactions in stars and the Sun a bit more energetic than they are in our universe, or a bit less energetic; or you alter the mass of the electron or its electric charge, or the violence of the original Big Bang. Any of these – get any of them out by even a small margin, and life won't happen.

Yet all these conditions were met in our actual universe. This extraordinary set of circumstances, making for the development of life against seemingly impossible odds, has been called the Anthropic Principle.

What are we to make of it? Are they just coincidences? If so, we're talking of odds far in excess of winning first prize in the national lottery. Or are we to conclude that physics has found God?

Those who think the latter are obviously moving into controversial territory. But Paul Davies is undaunted. I asked him how he accounted for the laws of nature being so fine-tuned for producing life.

'A number of colleagues accept this simply as a package that just happens to be. They will say, "Davies, why are you so worried about where these laws come from, or why they are those particular laws? Just accept the laws. Get on with the job." Well, when I was at school, I was always asking questions. And the answer I used to get was, "Sit down, Davies, and shut up!" Well, I'm *not* going to sit down and I'm not going to shut up. I'm going to continue to ask "Why, why, why", because, whereas many of my colleagues may feel comfortable with this astonishingly felicitous set of laws as just something that has no reason, it seems to me that the essence of the scientific method is to regard the Universe as having a rational basis, and arbitrary or absurd features are not really allowed. To stop the line of questioning at this stage seems to be far too premature. Our existence as sentient beings in the Universe links in to the basic laws of the Universe in a very meaningful way. And I think that provides human beings, in a modest but nevertheless important way, with some sort of deeper significance to their lives.'

Is there some simple way of accounting for the Anthropic Principle? Peter Atkins has a suggestion that allows him to explain away all the apparent coincidences at a stroke. He does this by raising the possibility of there being additional universes to the one that we find ourselves in (a suggestion which incidentally runs counter to his earlier proposal that there might be only one kind of conceivable universe).

'It is possible that this is not the only universe,' he told me. 'It's possible that universes are falling into existence while we're speaking at the moment. Somewhere there are trillions and trillions of universes. And it's possible that all those universes have a different mix of fundamental constants. Some have $\pi = 2$, others have electrons the size of elephants, and so on. Some will give rise to matter, but not to life. Others won't even give rise to matter – they will be just another boring universe filled with radiation. You can imagine a whole crowd of billions and billions of universes, and it just happens that one of those (maybe more than one, but at least one of those) happened to tumble into existence with a particular mix of fundamental constants that allowed life to develop.'

'So it was a kind of freak universe. And because we are ourselves a form of life, we have to be in one of the freak universes.'

'We ourselves can't avoid being in a freak universe,' he said. 'The consequences of the fundamental constants were first of all to produce galaxies, then planets, then us. And who knows what follows us? So I don't think one is forced to the view that there has been an active creator working to a design in the background. There are far simpler ways of accounting for the apparent fine-tuning of this Universe.'

'In what sense are you using the phrase "a simpler way"?' I asked. 'Some people would think that to postulate an infinite number of universes – all run according to different laws of physics – was hardly simple!'

'I think a "God" is really the embodiment of infinite complexity. Although the phrase "God created it" is extremely simple, it's actually a fraud. It conceals inside it activities of extreme complexity, a God working furiously to organize a universe with all its mix of fundamental constants. I think it is much simpler to say that universes just tumble into being (whatever that means), and if one of them happens to have the right mix, then life will take hold in it. This is intrinsically much simpler than a designed universe. So even though it might be very profligate (having a trillion trillion useless universes), this is still very much simpler than having a deity taking the trouble to design one.'

'I agree that the postulate of a "God" is a totally different kind of postulate – a fresh unknown – whereas with your idea, you're just having more of the same. But surely for somebody *who already believes in God on other grounds* – they feel that they have already contacted God through their prayer life, say – then surely for such a person they're not postulating the existence of anything other than what they already accept?'

'Well, they're deluding themselves,' Peter declared emphatically. 'I don't believe there is any argument in favour of that view. The fact that people have, for external reasons, decided that there is a God doesn't give any force to the argument that there is a God from the point of view of science. Science can show that everything could happen without needing to invoke the existence of a God. So the fact that some people have hallucinations; that some people feel so uncomfortable about the prospect of their own annihilation that they invent gods to give them succour; that some people invent gods in order to exert power over the less fortunate – none of this adds anything to a scientist's view that the world can come into existence without the intervention and the design element that is normally implied by deity.'

By now, Peter was well into the swing of one of his favourite activities: religion-bashing! It wasn't exactly relevant to the topic in hand, but I couldn't help rising to the bait …

'You talk about people inventing gods for succour and comfort, and for gaining power over other people, and I'm sure something of that sort does happen,' I agreed. 'But if one goes to the heart of a religion like Christianity, one's then talking about a religion which is very challenging, very self-denying. Now why would people want to invent a religion of that kind? If you're going to invent a God, why not invent one that was more comfortable to live with?'

'My suspicion is that religions get themselves invented in order to keep their priests in power,' Peter countered. 'If I can say to someone, "Look, do as I say – if you don't, then you're going to be punished for evermore," now I've got power. I think religions have grown as an exercise in power, and that's all. They have preyed (and that's not meant to be a pun!) – they've preyed on the beliefs of people who are in fear of the prospects of an after-life.'

'I concede there's a good deal of truth in what you're saying, Peter. Obviously the Church is a very human institution, subject to human frailty. When one looks back over its history, yes, there have been some very power-hungry people at the head of the Church. And yes, they certainly did try to frighten people into a belief. But if you go back to the founder of Christianity – to Jesus himself – do you really think that Jesus was putting forward a kind of message, the main purpose of which was for him to exert power over people?'

'Well, who knows what message this person Jesus was actually trying to get across? All we have in the Bible is a political manifesto, largely.'

'You paint a picture where the main motivation of religious believers is fear of the consequences of not being religious. But that's not the case for myself. On a Sunday morning I don't get up out of a nice warm bed and go to church because I'm frightened of going to Hell if I don't. I get up out of a sense of love and loyalty. And I'm sure that goes for the vast majority of the other people who go to church. Where does the sense of joyous love which so many religious people manifest come into this equation?'

'Well, it's certainly the case that people who believe seem to be much nicer people on the whole than people like myself who don't believe. We tend to come across as rather arrogant in our attitude towards the world. But I don't think that a distinction between being nice and being arrogant should be used as a criterion of who is *right*. I believe that science is the right way of understanding the world, and although other people might be much nicer than me, and go around helping people in slums and thanking God for their daily bread, or whatever, that doesn't mean to say that they're any closer to telling us the truth about the structure, the mechanism of the world.

'There's no real validity for belief in God, in my view. It's just a state of the brain that people are driven to by their upbringing, their conditioning, and their fears ...'

Discussions with Peter Atkins are guaranteed to be lively and stimulating, but I suspect they sometimes generate more heat than light! So let us get back to our original topic: the Anthropic Principle.

We have heard how Peter regards an infinite number of universes, all run according to different laws of nature, as being more simple than the notion of a 'God'. Paul Davies disagrees; he believes that the simple solution to the Anthropic Principle is that there is only the one universe, and that it may be the way it is for a reason. But would he go so far as to say that this happy set of circumstances had been designed by someone, namely a God?

'It's certainly consistent with that. This is really a question of your threshold of conviction. As the philosopher John Lesley has remarked, if every time we turned a rock over we saw the message *Made by God* stamped on it, then I guess everybody would have to assume that we did live in a universe of his design. It has to be a matter of personal taste whether you regard the accumulated evidence as compelling enough to want to make that leap. But inevitably it's outside the scope of science as such. Science deals with the facts of the world, religion deals with the interpretation of those facts.'

And of course, not only is the idea of God beyond the scope of scientific verification, but also Peter Atkins' idea of other universes – other universes being by definition not accessible from this universe.

I next spoke to Sir Hermann Bondi, mathematical physicist. (He looks so young and sprightly that I had to ask him whether he had yet retired. 'Retired!' he exploded with a laugh. 'I retired when I was sixty-five; I retired again when I was seventy; and I have just retired yet again at seventy-five!') He is President of the British Humanist Association and an agnostic. So did he see any significance in these so-called coincidences?

'Many pieces of knowledge in science that originally looked unconnected – "coincidence", if you like – later found some sort of explanation. So to speak of a coincidence as something that necessarily and for all time will remain a coincidence, I would be concerned about making such an emphatic statement.'

'So are you pinning your hopes on the science of the future being able to explain away these coincidences? It will be no mystery that our universe seems so well-designed, or set up, for life to come into being?'

'I am confident there will always be plenty of mystery! We live on a small island of knowledge in a huge sea of ignorance. What we do in science is to reclaim some land for that island. But of course the new island is still surrounded by ignorance.'

'Some people suggest that perhaps one way round all these coincidences is to say that there might be an infinite number of universes all run on different lines. Do you have any particular view about that suggestion?'

'To me, science is that which can be tested by experiment or observation. The assertion of many different kinds of universes leaves me stone cold.'

I asked him whether he saw any signs of intelligent design behind the Universe, behind the operation of the laws. He replied that people have always made arguments for the existence of God from design. The best-known of these was based on the fact that the various parts of the bodies of living creatures were so well suited to fulfil their respective functions, it seemed only natural to conclude that Someone must have designed them for those purposes. He went on: 'Quite a few of these arguments are no longer fashionable because there's now a good way of accounting for them. I am not convinced that the currently considered arguments from design have any more durability.'

'What about a rather different kind of argument: the world is intelligible. Would you say that was an argument for some kind of Intelligence as being the root cause of such a universe?'

'I would like to quote Einstein, who said, "The most unintelligible thing about the Universe is that it is intelligible." I'm not excluding the possibility that we will be driven to accepting an intelligent design, but I must say I find the present arguments rather unconvincing.'

'In any case, there would still be quite a gap to bridge from intelligent design to an Intelligence that had an interest in *us*,' I suggested.

'Oh, that's a huge step. I think there are two huge steps there. First, as you put it so very well, the huge step from intelligent overall design to interest in individuals, and then another huge step from that to any historical revelation.'

In seeking reactions to the Anthropic Principle, we have so far heard only from scientists. How about theologians? Nancey Murphy lectures at the Fuller Theological Seminary at

Pasadena, California. Like Ian Barbour and myself, she is an adviser to the John Templeton Foundation, so I was able to speak with her at our meeting in Atlanta. How did she respond to the coincidences?

'I think they're fascinating,' she said. 'The very fact that it has led cosmologists, who weren't in the least bit interested in religion, to raise questions about whether the Universe had a designer, is a very significant fact in recent intellectual history. Of course, you can't prove that God exists using those coincidences; there's always the possibility that they can be explained in some other way. But just as we've had arguments in the past that helped us to explain what we mean by saying that the Universe was created, and that it has a purpose, so there are new facts that we can use in our own scientific day and age to make those same points.'

Roger Trigg, professor of philosophy at Warwick University, also responds positively. 'There is a very real sense in which the Universe (as I remember one physicist said) "knew we were coming". That the whole thing was – I hesitate to say "designed" – but certainly it happened in a way that made it possible for life to be produced. Life isn't just an unimportant by-product in a very minor part of a minor solar system, in a minor galaxy. It is part of the very constituents of the Universe.'

'How would you account for this "design"?'

'I think that it does point to something like an argument from design. It is a modern argument from design for the existence of God. Now I know that it isn't a knock-down argument; other people may see it differently. Some people talk about an immense number of universes and it just so happens that we're in the universe that's produced us – we wouldn't be in one that hasn't produced us! But I think if the answer to a question is an infinite number of universes, one's in great difficulties. I think it's much simpler to believe in God who created the one universe, rather than saying there are an enormous number and we just happen to be in the one that's come up in this way.'

*

Sir Hermann Bondi spoke of the way previous arguments from design had later succumbed to natural scientific explanation. As we shall be seeing in the next chapter, the argument based on the 'design' of the bodies of living creatures – seen as a knock-down proof of God's existence – came to grief over Darwin's Theory of Evolution by Natural Selection in the last century.

Now, in our own times, religious believers seem to have been handed a second argument from design, this time from physics and cosmology rather than biology. Should they use it to try and convince the sceptical? Bob Russell, a former physicist and Founder Director of the Center for Theology and the Natural Sciences at Berkeley, California, sounds a warning about how one might use the Anthropic Principle. 'If we don't know our history, we will relive it – and be burned by it again. There is no theology through science; there is no direct proof of God, full stop! I think you can say, through the Anthropic Principle, that there is stronger supporting evidence, or stronger supporting reasons, for a theistic view.'

'So are you saying that if you already believe in God on other grounds, the Anthropic Principle is confirming for you that that is the right path to go down?'

'I think that's a good way to put it. It doesn't *prove* God, but it certainly tends to support and confirm and strengthen (like a character witness in a case) the idea that the theistic claim is the best possible explanation. It's more explanatory, more fertile; it can account for more things than the atheistic one. Ultimately, the atheistic one is still posed with the question "*Why* is there a universe"?'

Chapter 2

Life

Then the Lord God formed man of the dust of the ground, and breathed into his nostrils the breath of life, and the man became a living being ... The Lord God caused a deep sleep to fall upon the man, and he slept. Then he took one of his ribs, and closed up its place with flesh. And the rib which the Lord God had taken from the man he made into a woman.

The familiar words of the Genesis account of human origins. But a far cry from the view of modern science ...

Evolution by natural selection

We have seen how the raw materials for life are scattered throughout the Universe. Some of these have been incorporated into the planets. Some of the planets (like the Earth) are to be found at such a distance from their star that their surface temperature is neither too hot nor too cold to sustain life. But how does one go from the basic raw chemicals to ourselves and to the other living creatures we see about us today?

In the UK there is probably no one better able to explain the theory of evolution to the non-expert than zoologist Richard Dawkins. I was invited round to his Oxford home – a remarkable place, filled with old fairground animals he and his family have collected from disused merry-go-rounds. Having thrown my raincoat over the back of a wooden lion just inside the door, I settled down on the settee in his lounge, surrounded by several horses. I asked him how he saw us and the animals (the real animals, I mean) coming into being. What does Evolution by Natural Selection entail?

'Evolution is not necessarily tied to natural selection,' he explained. 'You can have a theory of evolution that's not natural selection. Natural selection is Darwin's particular theory for the *mechanism* of evolution.

'Evolution itself means that species are descended from other species that are different from themselves. So we are descended from something like an ape, and before that from some kind of insectivorous small mammal. Before that from a reptile, before that from a fish, and so on – back to the earliest of our ancestors, which would have been something like bacteria. That's evolution itself.

'Natural selection, which is the main mechanism of evolution, is the non-random survival of randomly varying hereditary characteristics. That's putting it in one sentence. What it means in practice is that within any one species, animals vary. Some of them are taller, shorter, faster, slower. At least a part of that variation is inherited – it's genetic –and part influences whether or not you survive to reproduce. So if we're talking about, say, cheetahs chasing antelopes, those cheetahs that can run fastest will tend to be the ones that catch the antelopes, and will therefore be the ones who have children and pass on the genes for running fast. And that's why cheetahs run fast. And that argument holds for all living creatures – why they have such good eyesight, why they have hands that can grasp (in the case of monkeys), why they have hooves that can gallop (in the case of horses). Everything about an animal is shaped by this non-random survival, owing to the fact that some hereditary characteristics are good for survival and reproduction, and others not. So natural selection is the explanation of why evolution has gone in the direction of apparent design.'

'And how far back does this go?'

'It goes back about three billion years – that's 3,000 million years or a bit more. But for about the first two billion years of life, things would have looked a bit dull from our point of view. There would have been mostly bacterial life. And then about one billion years ago, it all started to get more interesting and you got multi-cellular life (life that's built up of many cells), which therefore starts to be big – big in the sense that you and I can actually *see* it. You start to get animals and plants that have a kind of shape and form that you can see without a microscope. Then it really took off, and you started to get the sorts of large, complicated creatures that we see around us now.'

'And all this goes back to inanimate chemicals?' I asked.

'Before life itself began it had to go back to inanimate chemicals. The key event for starting off this process of natural selection is the arising of a chemical that is capable of making copies of itself. That's the key point: a *self-replication* event had to take place. Once you've got self-replication, you then have the possibility of an exponential increase in numbers. You have two, and then each of the two makes two more (that's four), then eight, sixteen, thirty-two, and so on. An additional ingredient that's necessary is some variation. It mustn't be just one kind of molecule that's expanding in numbers; there's got to be at least two (and preferably an indefinitely open-ended variation), so that if there are two different kinds of molecule, then one of them will happen to be slightly better at making copies of itself than another. That's the beginnings of natural selection.'

'And what's the cause of the variations? Is it mistakes?'

'It's mistakes, random mistakes. And to this day, all the variation that natural selection works on, even in ourselves, is based upon random mistakes in the molecules which we call genes, the DNA molecules.'

'So in this way you go from inanimate chemicals, which clearly are not alive, to ourselves who clearly are alive,' I said. 'What do you see as constituting life? What actually is the difference between something that is living and something that is non-living?'

'"Life" is one of those words that I prefer to draw lines around. It's a bit like if you ask me, "What constitutes a tall man?" Does he have to be six foot or six foot five? I don't really care exactly when a man becomes "tall". I can see that people are relatively taller. And it's the same way with life: I don't want to be pushed into the position of saying, "At this point life began". One might be tempted to say life began when the first self-replicating chemical came into existence. But I think I'd prefer to say that life gradually came into existence. And now we have a situation where there's no question about it – we have life.'

'So it's really a question of degree,' I suggested, 'rather than being something qualitatively different?'

'Yes. I think it's a question of degree.'

Evolution first came to the public's attention through Darwin's famous book *The Origin of Species*, published in 1859. He wrote it at his home in the village of Downe in Kent. The house is now preserved as a museum.

Darwin's study is lined with books and drawers of specimens. His armchair is a battered old thing covered in black leather, mounted on an iron frame and supported on wheels. Next to this chair is a walking-stick. It was through using this as a long-handled hook that Darwin was able to yank himself around the room to retrieve books without the need to get up out of the chair. To my delight I was given permission by the curator to sit in the chair – a privilege, I hasten to add, not normally accorded to visitors. I balanced a wooden board, covered in faded cloth, on the arms of the chair so that it was positioned across my lap. In the past, the repeated placing of the board like this had been responsible for wearing the leather on the arms right down to the stuffing underneath. It was in this chair, and on this very board, that Darwin had written that momentous book, destined to change for ever humankind's understanding of itself in relation to the rest of the animal kingdom.

My main reason for being at Down House was that I had arranged to meet Sam Berry, professor of genetics at University College London. I asked him how *The Origin of Species* was first received.

'It changed people's thinking in a major way,' he said. 'During the first half of the last century, there had been a lot of strains put on the traditional understanding of how the world came into being – the Genesis creation stories. It was increasingly apparent that the world was older than 4004 BC, which was the accepted date. It was apparent that some species had become extinct, which didn't seem proper if God had created everything perfect. By the middle of the last century the outlines of the fossil record were pretty well known, but still the notion of biological change hadn't become accepted, because there was no mechanism known.'

'Is there not a worry about the evolution of very complicated structures like, for example, the eye? Wasn't Darwin himself a little concerned about something as complicated as the eye, and how that could have arisen in this random way?'

'Darwin talks about the evolution of the eye in *The Origin of Species*. You have primitive organisms which have a light-sensitive spot. Now if you can tell where the light is coming from, this may be of great advantage to you. That spot tended to get down into a hollow, and then it began to be focused, and then you could move it. All these stages can be recognized in animals, leading stage by stage up to the eye as we know it.

'In no way can we think of an eye just appearing out of the blue, with a single mutation. Half an eye would be no use. But all these primitive eyes were of considerable advantage. They were selected in the Darwinian natural selection way. This is speculation, but it's very reasonable speculation. So there's no problem about these "organs of perfection", as they are often called.'

'There are still quite a number of gaps in the fossil record,' I pointed out. 'In view of these, would you say the theory of evolution is proved beyond doubt? Can you explain those gaps?'

'Before I talk about the gaps, let me just explain one thing which people very often get confused about when talking about the "theory" of evolution. "Theory" is, to most people, the theory of a detective story: you have a theory about who did the murder. "Theory" in the scientific sense has a much more precise meaning: it's an idea *which has been tested*; it has been tried out with all sorts of experiments and is much more than just a mere idea. So the "theory" of evolution is really a fairly central part of biological understanding.

'Now, is it acceptable to normal thinking beings? Yes, I think it is. The fossil record has gaps in it. But you would expect there to be gaps, as Darwin himself pointed out in *The Origin of Species*. Not all organisms are going to leave fossils, because the actual process of fossilization is a fairly specific one. It is rather remarkable that we know as much as we do about the fossil record. So in answer to your question: No, the whole problem about gaps in the fossil record is annoying, but it's not a real gap in the understanding of the theory.'

So in purely scientific terms, the theory of evolution is in good shape. But what was the original reaction to it from religious circles? Was the response as universally hostile as most people have been led to believe?

David Livingstone at Queen's University, Belfast, has made a study of Protestant history on both sides of the Atlantic. He told me, 'The initial response of Christian people to Darwinism was a complicated affair. For that reason, I think that many of the ways that people have thought about this – assuming, for example, that there was a *necessary* conflict – are perhaps rather too simplistic. Overall, amongst conservative Christians, there was rather perhaps less panic or conflict than has sometimes been imagined.'

'Then how come there is this modern understanding of what happened as being a conflict? How did that arise?'

'The idea of a conflict emerges from a couple of different sources. One is the writings of historians. Around the turn of the nineteenth century, a number of books with titles like *The Conflict between Science and Religion* were published. Sometimes people read the titles and didn't read the books. They generalized to an overall conflict, or warfare, between the two domains, without really reading the subtleties of what these historians were after.

'There's a second thing. In the early decades of the twentieth century, particularly with the rise of fundamentalism in the United States, there were, as you are aware, quite a number of dramatic court cases over the teaching of evolution in schools. Opposition from the fundamentalist quarter to the teaching of evolution engendered an image of conflict.'

'At the time of Darwin, were there any theologians who actually welcomed his findings?'

'It's difficult to know what we might mean by "welcoming his findings". Let me put it to you this way: there was a range of responses, from reasonably enthusiastic to rather less certain.

'Consider, for example, the writings of the outstanding Princeton theologian, Warfield. In 1916 he wrote a very influential article about John Calvin's doctrine of creation. In this article, he actually goes so far as to say that Calvin's doctrine turns out to be a pure evolutionary scheme. What he meant was that Calvin believed God had really only intervened in the creation process at two points: the initial act of creation, and then secondly for the creation of the human soul. Everything else had come about through the operation of what Calvin called "intrinsic forces in nature". For that reason, Warfield felt that it was really Calvin, and not Charles Darwin, who was the author of the theory of evolution.'

One of the things that has always intrigued me about the supposed early opposition of religious people to Darwin's ideas was the fact that, when Darwin died, he was accorded the considerable honour of being buried in Westminster Abbey. When at Down House, I had asked Sam Berry whether he saw any significance in this, given the fact that Darwin himself had always assumed he would be buried just along the road from his home, in the village of Downe.

'It was part of the very positive coming together of faith and science,' he said. 'Darwin destroyed the possibility of the viewpoint that God, having created the world and made it perfect, then retired above the bright blue sky. If there was going to be a viable belief, then God had to be *in* life, holding together life, acting through normal processes. In theological language, God was immanent as well as transcendent. So Darwin, in a sense, brought the two together, and in one way his burial in Westminster Abbey was a sign of that coming together.'

Coming back to the present time, I got to wondering how a place like the Natural History Museum in London handled the questions raised by evolution. I went to find out, and was met at the entrance by Neil Chalmers, the Museum's Director. We went up to the gallery overlooking the main hall. At the top of the grand flight of stairs we were confronted by a notice-board announcing *Our Living Relatives*. It mentioned things like gorillas, gibbons and orang-utans.

'This is your set of exhibits about evolution by natural selection,' I said. 'I've already had a look through it. It seems to explain things very well. But I have one niggling worry. People coming to this exhibit will bring with them preconceived notions about human origins. For example, some of those school parties down there in the hall may very well have come from a school assembly this morning where they had a reading of the Adam and Eve story from Genesis. Then they come here, and it's a totally different kind of world. Now I can understand you not wanting to get into the thorny business of how to interpret the Bible, but don't you see a kind of educational problem when people come here with preconceived notions and you're not addressing them?'

'I think we have to start by asking who our visitors are,' he said, 'and where they come from. The fact is that we get visitors from all around the world, from a whole range of different religious and ethnic backgrounds. It would be impossible, and indeed wrong, for us to try and anticipate what those religious beliefs and backgrounds were. And so I think we have to start from our position of strength as a scientific organization and say that science can tell us this about human evolution. And then it is up to the individual, and the religious community, to put that information into context in terms of their own beliefs.'

That was fair enough. But what *are* the options for religious people when it comes to the Genesis story of Adam and Eve? Is a literal approach to it the only one? If so, confrontation is inevitable. Some people think that that is exactly the way it is. Richard Dawkins lines himself up firmly on the side of evolution, and sees no need at all to bring in the idea of a creator god.

'I call it explanatory overkill. It's putting two explanations in where one will do. The theory of evolution by natural selection is on its own sufficient to explain life. It may be that God on his own is also sufficient to explain life. If I were God I wouldn't do it by evolution! I would do it directly. By invoking the idea of evolution by natural selection as God's way of doing it, you are in effect invoking the one way which makes it look as though God isn't there. So if God chose that way of doing it, then he deliberately chose a way which totally covered his tracks.'

'If he *was* there, and this was in fact the way he did it,' I persisted, 'would you say that the nature of this particular process casts some light on the kind of God he would be?'

'I think it would show him to be totally indifferent. It wouldn't show him to be positively cruel, as some might think, nor would it show him to be positively good. Nature has no pity. The antelope probably suffers horribly when it's being chased by a cheetah and caught by it, and the suffering of antelopes being chased by carnivores is probably nothing compared with the suffering that goes on all through the world every day with parasites gnawing your vitals from inside. The consequence of natural selection is suffering on an enormous scale all over the world. It's not that nature is malevolent, it's not that nature has it in for us and is trying to make life miserable. It's just that misery of this kind is precisely what you'd expect if nature is totally indifferent to suffering, and is only concerned that each individual is maximizing the survival of its DNA. So what I see in the Universe is nothing but pitiless indifference. And that looks to me like no God. If God is there, then he is neither bad nor good; he's indifferent.'

We heard earlier how David Roseveare, representing the creation scientists, was opposed to Big Bang cosmology. As might be expected, he has no time for evolution either. I asked him how he could possibly reject what the vast majority of scientists are agreed upon.

'Let me first say that I myself am a scientist. The vast majority of scientists believe in evolution because they believe that the vast majority of scientists believe in evolution! There are just a small number of scientists who actually work in this field.

'My own belief is that the Universe, and life itself, show distinct signs of being carefully designed. The Bible in fact says, "In the beginning God created the heavens and the Earth". The Lord Jesus Christ, who for Christians is the creator, spoke of Genesis 1 in a *factual* way. He himself regarded it literally. He said, "Have you not heard that He that made them at the beginning, made them male and female", when asked about divorce. He referred to himself as "the Lord of the Sabbath". As you know, the Sabbath is the fourth commandment, which underlines the idea that God created things in six days.'

'Yes, but there are many people who accept that God created the Universe and that he created human beings, but that he did it through the processes uncovered by modern science,' I protested. 'Why do you think that Genesis has to be taken as a *scientific* account of our origins? What makes you think that the original authors of Genesis intended it to be read in that way?'

'Simply because the writers of the New Testament refer to Genesis in a literal way, and the Lord Jesus himself referred to Genesis literally – to Adam and Eve. For instance, Paul says, "As in Adam all die, so in Christ we shall all be made alive".'

I continued, 'The idea that Genesis ought not to be taken as literally true – it's an example of an ancient story which conveys timeless spiritual truths about ourselves, but in story form – is not new. It hasn't come about as a result of modern science. If you go back to ancient times you have, for example, Gregory of Nyssa saying, "What man of sense would believe that there *could* have been a first, and a second, and a third day of creation, each with a morning and an evening, before the Sun had been created?" He was speaking in something like the fourth or fifth century. You have St Augustine saying, "In the beginning were created only the germs or causes of the forms of life, which were afterwards to be developed in gradual course." That is not the theory of evolution by natural selection; he did not know about the mechanism of natural selection as such. But it was certainly an evolutionary idea of some kind. And St Augustine was one of the greatest Christian teachers of all times. He was speaking, again, around 1400 years before Darwin. So what makes you think that the literal interpretation of Genesis is the one which the early Church accepted, when these great leaders clearly did not go along with that viewpoint themselves?'

'Evolution itself is a very old view. But as far as the Christian Church is concerned, it was more usual for them to believe that the account in Genesis should be taken literally, rather than be interpreted in terms of six long periods of time. I come back to the New Testament writers who believed it like that. I would say that if you look at the language of it, there was quite clearly evening and morning at day one, etc. And the fourth commandment likens the days of the week to the days of creation: "Six days shalt thou labour, for in six days the Lord made the heavens and the Earth".'

As we have noted, the creationist movement is particularly powerful in the United States; it is a force to be reckoned with.

I asked Philip Hefner, Director of the Chicago Institute for Religion and Science, to describe the creationists' activities. He began: 'They insist that if scientific evolution is to be taught in the schools, then Genesis must be taught as well – because Genesis is an alternative form of what's covered by scientific evolution. The courts have ruled that this is really a religious claim, and not a scientific claim. But the creationists have never acknowledged that. They are just as vigorously fighting to get it into the school curricula as they've ever been, even though the courts have ruled against them.'

'How easy is it to put a figure on the number of Americans who hold such a view?'

'I am not sure it's so easy. There have been many polls taken, and some say that as many as half to two-thirds of the American public takes the Bible literally, particularly Genesis. I doubt that myself. But certainly thirty or forty million would be a low figure. It's substantial. Creation and evolution is a hot topic, even after a hundred years of debate.'

'What sort of demands do such people make on the school system?'

'Their demands rest primarily on "equal time" claims. They want the textbooks, if they talk about scientific creation, also to talk about Biblical creation.

'One of the underlying issues here is what's often called "materialism". The creationists are afraid that if evolution is accepted as the basic interpretation of the world – its origins and development – that means we have a closed materialistic system which eliminates God, and eliminates the spirit. So when they talk about putting Genesis into the curriculum, they are also talking about anti-materialism. That discussion rests on an outmoded understanding of nature. It insists that the scientific account is a one-dimensional, rather banal, materialistic account – which of course it isn't.

'Many high-school teachers would like to discuss these issues. They recognize the seriousness of them, and want to respect the value systems that the students bring with them to the class. But most teachers are afraid to discuss anything in depth, because they are afraid that they will be the object of attack from these creationist groups, and possibly even lawsuits.'

'So how much more difficult does this make it for those in the United States who do *not* see religion and science as being in conflict?'

'If I am dealing with the media, reporters always want to discuss creation and evolution. When we get into more complex issues, they lose interest. Further, they want to make it a warfare issue. Another difficulty has to do with our traditional separation of church and state (in the USA). It is virtually impossible to receive any public money for the kinds of research and studies that we're doing at the Chicago Institute for Science and Religion. I've received government money from Britain, Germany, South Africa and Denmark, but I could never get any public money in the United States. There is no difficulty when it comes to attracting people who want to discuss the issues in a sophisticated way. There are just as many people in the United States who want to go beyond the creationist tangles. But they do so without the support of public discourse or public finance.

'It's ironic that a church denominational seminary like my own is actually freer to pursue this discussion than most universities, because the universities have to face all these public constraints.'

*

Biologist and atheist Will Provine enjoys stoking up the creation-ism versus evolution controversy. He begins by complimenting the creationists: 'I think creation scientists are very intellectually hon-est in their beliefs. If evolution is true, then none of the things that deeply religious people want to be true are in fact true. No God. No life after death. No free will. No ultimate meaning in life and no ultimate foundation for ethics. All of these things are taken away, and I believe creationists have a keen appreciation of this fact. So I sympathize with their general point of view. In other words, they say evolution cannot have occurred. I understand the sentiment. I just believe they're *wrong*. If modern evolutionary biology is true, then the traditional foundations for religion are gone.'

I put it to him, as I did earlier to David Roseveare, that St Augustine subscribed to an evolutionary idea, of sorts, long before Darwin. And there are many religious people today who are happy to accept Darwin's account of our origins. How did he account for that?

'I understand how desirable a compatibilist view of religion and science would be,' he replied. 'Almost all of the religious traditions want to be up to date at the present time, that is, to encompass evo-lution or whatever else is true in the scientific world. The problem is, it seems to me, that an evolutionary explanation is just incom-patible with a Christian, or a Hindu, or any other kind of view that suggests purposive forces guided the process producing the organ-isms we see now. There is no evidence whatsoever in the natural world of any kind of purposive force working.'

Sam Berry, besides being a geneticist, is also President of the 750-strong Christians in Science movement. How did he view creation science?

'It's a very over-simplistic viewpoint. I first came up against it when a medical student friend of mine said to me, "You know, I could never become a Christian." And I asked why. He said, "Because it would be intellectually dishonest to give up believing in evolution." Well, I said, "Get lost! What has evolution got to do with being a Christian?" But to him it was either one thing or the other. To me, this is absolute nonsense.

'It was his response that started me thinking about what we are actually being told in the Bible. What we are told has to be reinterpreted in every generation. For example, we had to reinterpret what the Bible was talking about when it seemed to say that the Earth was so fixed that it couldn't be moved. Actually what the psalmist was talking about was the unchangeableness of God; he wasn't talking about astronomy. We can't learn our science from the Bible.

'I am an ape. I am an ape made in the image of God, and I've got to bring these together. This doesn't mean watering down either in any way. If there's any conflict, I've got to look at both my faith and my science. It could be that my science is awry; it could be that my faith is based on a wrong proposition.

'In Darwin's time,' he continued, 'it was necessary to go back to Genesis and see what the Bible was actually saying. We were told in Genesis that God made the world, that he progressed from the simple to the complex. We are not told anything about mechanisms. Science is concerned with mechanisms; faith is concerned with the "why" questions. The "how" and the "why"; science and faith. These are different aspects of life. If we get hung up on one, we are not able to be fully coherent about ourselves as real human beings. So many Christians have a half-baked faith; for them, it's faith or science. But it isn't that, it's faith *and* science.'

Which raises the question of what exactly we are being told in the Bible. I asked Ernest Lucas, former biochemist and now lecturer on the Bible, whether we should conclude that the story of Adam and Eve has been caught out, or is it tackling some other kind of question?

'It was asking questions such as "What does it mean to be human?", "What's my relationship to God, or the gods – if there are gods?", "Why do we have to die?" And it was seeking to give answers to those kind of questions in terms of stories – stories which have a profound meaning to them.'

'Why use story form rather than straightforward narrative?' I enquired.

'Well, it is a form of narrative, but one in which truth is put across in a symbolic way, because it was dealing with the big questions of life – the questions of meaning and purpose. I think these are more powerfully put across in story form. You're dealing with a culture where you haven't got schools and universities. In any culture before modern education, stories were a major form of communicating.'

'And this was common amongst ancient civilizations? It wasn't just the Israelites who used this kind of communication?'

'No. We now know quite a lot about Semitic culture in the ancient Near East in the time of the Hebrews – material from Babylon, also material from Egypt. They've each got their creation stories in which they give their answers to the questions "Why are we here?", "Why do we suffer?", "Why do we die?", "What's our relationship to the gods?" So it was the normal form of answering these kinds of question.'

I wanted to know how the Genesis creation stories compared with others that were in circulation at the time. Were there features in common? Was there something particularly distinctive about the ones that we have in the Bible? He explained: 'There are certainly some things in common. You can see similar symbols being used. For instance, the symbol of a tree of life, which you will find throughout the ancient Near East of that time. The symbol of a snake robbing humankind of immortality – that occurs in the Babylonian story.

'But you can equally see differences in emphasis. For example, in the case of "Why aren't we immortal?", the Babylonian story has a non-moral answer. Whereas in the Hebrew story, it's put very much into a moral framework: the creatures have rebelled against their creator; they've sought to become gods themselves.'

'What would you say to someone today to explain why they should continue to take the Adam and Eve story seriously? If it's not meant to be a rival to the theory of evolution, what other kinds of truth are there in it?'

'You read the story in Genesis 3 about the Garden of Eden, and Adam and Eve's rebellion against the Creator. You find in that story a very deep truth. Once their relationship with their Creator has been broken, three *other* relationships go wrong.

'The relationship of Adam with himself (and Eve with herself) goes wrong. They feel shame and guilt, and they hide. There I think we have a deep insight that psychological problems arise when we are trying to be gods ourselves, and are not relating to the Creator outside of ourselves.

'Then the relationship between Adam and Eve goes wrong. When God says, "Why have you done this?" Adam blames Eve. He says, "Well, she gave it to me." So social relationships begin to go wrong.

'And then there's the profound truth that their relationship with the non-human creation goes wrong. When God talks about you having to struggle for your bread "in the toil and the sweat of your face", that relationship, which was meant to be harmonious, goes wrong. There is a deep insight there which is 3,000 years old. We might have got on rather better if we'd recognized it earlier in our dealings with the ecological crisis. It's not something that just needs a technological fix. It goes deeper – to our understanding of who we are, and how we relate to the Creator.'

'How about that curious idea that Eve was made from a rib taken from Adam's side? What sort of truth do you see in that?' I asked.

'It is saying that she is made out of the same stuff as Adam. Men and women have the same fundamental nature, the same origin, and they are equals ...'

'Was that a new idea in those times? Was man's superiority taken for granted?'

'I think the superiority of the male was largely taken for granted.'

'Would you say that there's the implication that man is not complete without woman, and woman is not complete without man?'

'Very much so. In Chapter 1 God says at the end of six days that everything is good. Then suddenly in Chapter 2 you get the statement that God says it is *not* good. What does he say is not good?

That man should be alone. There is something missing in his creation until woman comes to be the equal partner, and to share with man.'

'I'm right in saying, am I not, that there are in fact *two* creation stories here? The first chapter of Genesis, and then the second one, beginning at verse four of the second chapter. How come there are two of them, and what's the distinction between them?'

'Yes. Most Old Testament scholars do feel that we've got two different stories here. They are written in a very different style. Most scholars think that the story in Chapters 2 and 3 was written earlier.'

'So why do you think that the editor (if you can call him that) of Genesis included *both* of them? Is there something distinctive about them? Does the first chapter add something that we don't get in the second?'

'Yes. The first one paints a much broader canvas. It's concerned with the creation of the heavens and the Earth, of the whole cosmos (as it would have been known to the ancient Near Eastern people). And the stories in Chapters 2 and 3 have the much narrower focus of the origins of humankind, and their relationship with their Creator and with one another. So Chapter 1 provides the broader framework into which the story in Chapters 2 and 3 fits very nicely, picking up from the final act of creation: God created male and female in his own image.'

'You've been saying that one ought not to look upon these creation stories as a scientific account of our origins, but there are, of course, many people who do precisely that. Where do you think this literal interpretation of the Genesis story comes from? Is it the original interpretation, or is it something which has arisen since?'

'I don't think it's the original interpretation. It's asking questions that couldn't have been asked 3,000 years ago. The whole approach wouldn't have been of concern to people 3,000 years ago.

'I think it arises first of all because we have lost our sensitivity to the way literature can be used to express truth in ways other than literalistic prose. So much of our education trains us to look at things in terms of literalistic prose. I certainly found a great shift when I moved from scientific research to theological study. Through being able to study literary criticism, a whole new world opened up to me – a world of meanings that I hadn't recognized before.

'And I think it also arises out of the situation in the Southern States at the early part of this century. There were scientists who were saying the Bible is wrong in the light of evolution and so on. In an attempt to defend the Bible, some Christians felt they had to take it literally, because that seemed the only sure way to hold on to it. The only reliable way of expressing truth is the way scientists do it. But then they found that they had a head-on clash with science.

'I think it's a fundamentally mistaken approach. It is treating the early chapters of Genesis as the kind of literature it isn't, and never was. If you ask silly questions, you get silly answers. And I'm afraid it's asking the wrong question (and therefore a silly question) of the Bible to say, "What does it say about evolution?" It's not talking on that level.'

I next asked Phil Hefner whether he felt that his understanding of current science took away any of the value of the Bible.

'No, not at all,' he declared. 'I recognize that the Bible is largely myth. Of course, for many people the word "myth" means "it's a fairy tale, it's not true". But myth is the kind of language that talks about those dimensions of life that are difficult, if not impossible, to talk about in ordinary language: Does my life have meaning? Does my life have purpose? What does my freedom amount to?

'I find the Bible to be an extraordinarily rich resource: the figure of Jesus, the teachings of Jesus, the life of Jesus and his dedication to service and love of the world around him. I do not find these in the least lessened in the scientific world, whether you're talking about the environment, about Yugoslavia, American cities, the pluses and minuses of free-enterprise economics, or how we are to relate in a vast, global, multicultural human community. Jesus (and I wouldn't want to minimize the great figures of the other religions) is as relevant today as he ever was.'

*

So much for creation science and alternative ways of interpreting the Bible. It is one of the claims of religion that humans have a special place in nature; we are part of a divine plan. But, I wondered, in the light of evolution does that still hold? Does it any longer make sense to think of ourselves as distinctive, set apart from other animals? John Habgood sees no conflict in accepting the view of humans as distinctive, and at the same time as a product of evolution. I asked him, 'If we are to see ourselves as evolved animals stretching right back to a primordial slime of inanimate chemicals, presumably consciousness must have come in somewhere along the line. How do you see the dividing line between humans and inanimate chemicals?'

'Does there have to be a totally clear dividing line?' he ventured. 'It seems to me that nature shows us that things are continuous. But just because they are continuous, that doesn't mean they're all the same. For example, where's the dividing line between night and day?

'In evolutionary terms, one can see a gradually growing awareness developed through language. In language, you first develop the capacity to be aware of things, to think about them, and to handle them. This is a first step on the way to becoming aware that there may be some transcendent realities to which you can relate. So people talk about the world as a "vale of soul-making". In a sense it is. The whole process leads to what we now enjoy: the capacity to relate to that which totally transcends us, and indeed is our source and origin – namely, God.'

'What is the theory of evolution saying about ourselves and about God and his creative work – the way he goes about it?'

'One of the main things it did was to introduce a time dimension into ideas about creation. It was not just something that happened at a moment in the past. God is involved in a continuous process of which we are a part. The very method of creation almost allows things to make themselves, to develop themselves with a great degree of openness. This speaks to me of a God who wants to create a universe which, in some sense, is allowed to be itself. It is at some distance from himself *in order that it may respond to him.*'

*

Such is the view of a theologian. But how does a working biologist see himself in the light of evolution? Steven Rose works on the brain, and has a particular interest in trying to understand memory. As we talked in his office at the Open University, I found it a little disconcerting to have a pickled human brain floating in a glass tank by my elbow. I asked him how the theory of evolution by natural selection affected the way he saw human beings. Did he see anything distinctive about us? If so, did that have anything to do with our brains?

'First of all,' he began, 'let me say something about what you call the theory of evolution by natural selection. For me, and for most biologists, evolution isn't a theory; it's a fact. What is theoretical, if you like, is the mechanism by which evolution has occurred. Darwinian natural selection is one major mechanism, but it's not the only way in which evolution has occurred. There are a whole lot of other ways as well, and I think it is important to clarify that.

'But clearly, humans are the product of one particular sequence in evolution. Of that particular evolutionary sequence, we are the organisms with the biggest, the most developed, the most complex brains. What's distinctive about us, of course, is our capacity for language, our capacity for social activity. Once those capacities were with us, we could record our own history, develop tools, create and modify our environment, and learn not only from our own personal history, but from our social history as a species – in a way that no other animal species that we know of on Earth has been able to do. So that's what makes us distinctive, but we are distinctive as a particular sort of animal species.'

'Would you say it's a difference of degree or a qualitative difference?'

'Well, I think there are qualitative differences between many species. If there weren't those differences, then the species would not be distinct. (They are distinct in that they can't interbreed.) It's clear that there are unique features about humans. There is some rather disputed evidence that chimpanzees can be taught, slowly and painfully, the equivalent of a sign-language process. Some animal species can also use certain artefacts that you could call "tools". But the difference is so great between those and humans that I think we are dealing with *qualitative* differences.'

'When you think about our evolutionary origins, does that reduce your sense of awe and wonder at what it is to be a human being?'

'I don't think it reduces my sense of wonder at all. I regard the living world as a fascinating, extraordinarily beautiful, exciting domain – one which is incredibly interesting to study as a re-searcher. The reason why, as a biologist, I chose to work on the brain and on memory was that this seemed to me to be the last great unsolved terrain of biological problems. I am full of awe and wonder about that, but I do believe it's a scientific problem which is soluble – if not by the classical techniques of science, then by techniques which science can develop and *is* developing.'

Geneticist Sam Berry sees the difference between humans and the other animals in these terms: 'If you read the creation stories in Genesis, you are told that humans were made in the image of God. In Genesis 2 we are told that God took the dust of the Earth – something he'd made already – and breathed into it. It was a two-stage process. This is what I see as the difference between animals and human beings: it's God in us, the image of God in us.'

'When do you see this as actually happening, given that we have a smooth transition from our earlier ancestors? Do you see any well-defined position along that line where God breathes something of himself into us, and we then become something which is quite distinctive?'

'As a scientist, an archaeologist or an anthropologist, I see nothing. If you look at the fossil history of human beings – and the fossil history of humans is a lot better known than that of many other species – there is no discontinuity. I wouldn't have expected that. The image of God is nothing anatomical; it's nothing physiological; it's nothing genetical. It's the ability to relate to God. When did it happen? I may sound almost fundamentalist here, but I am taking the Bible records as the basis of my understanding. Adam was a farmer. He had a garden, he was told to till it and tend it. In archaeological terms, this means he was Neolithic – something of the order of, let's say, 10,000 years ago. Perhaps that is the time when God as it were took an individual, breathed into that ape (call it an "ape"), and that thing became a human being in the fuller sense of the word. So I put man and woman fairly late in historic time.'

'Would it more or less coincide with the development of language, and the sharing of experience between one person and another?' I asked.

'Yes. But I wouldn't like to tie it down too closely to that. Obviously the ability to communicate is part of our humanness. But the ability to relate to God, and the ability to relate to other members of the same species, are different things. So yes, they are of the same order, but no, they are not the same thing.'

For Arthur Peacocke, who is both a theologian and a former biochemist, the very nature of natural selection, the way it involves an element of randomness, is an essential feature of the creative process adopted by God: 'I think the theory of evolution has articulated, unravelled and made clear to us how – to put it theologically – God has been creating life and different forms of life. The evolutionary process is one which enables new forms of life to come into existence. But it does not answer the question *why* should there be such a process at all.

'God, in my view, gives existence to these processes. The processes are not just random, they are random events acting within a framework – a law-like framework which limits the possibilities. This interplay of chance and a law-like framework is, together, *creative*. If it was just chance by itself, then nothing would ever be stable enough to live long. If it was all law, you would have a mechanism like a clock. But the interplay of chance and law shows you a system in which structures can evolve and become something new. So it is actually a very subtle combination which allows a system to be both structured and progressive and evolving. So to me it's really a rather wonderful way in which the various potentialities imprinted in the Universe can be gradually made actual and real.

'The dice are loaded towards producing complexity. Complex forms have the sensitivity to record information from the environment. And so there is a propensity towards consciousness, because consciousness helps creatures to survive.'

Theologian Keith Ward also welcomes the randomness involved in evolution.

'If there is an element of randomness, it lets in some freedom, some openness, some creativity. But it's not going to be so big as to cancel out God's ability to say where the Universe is going, and God's ability to ensure that it gets there. So I'm all in favour of a bit of randomness.

'Evolution provides one of the most exciting new perspectives on God, and on human life, that there has ever been. It's far from being the case that evolution is against religion. It's very much something which gives a new insight into the way that God develops new creative things out of old and less creative things. Out of inert matter, banging around, you get rational, moral beings like us. And no doubt the world will go on evolving in ways we can't imagine. You can easily see the Universe, from an evolutionary point of view, as a movement from the material to the spiritual.

'Now Charles Darwin probably didn't see it like that, and this may be why some people have problems. They see evolution as a purely random or a purely materialistic thing. But I think you should see it as a *spiritual* development, the emergence of new qualities. And in that sense, it's rather a new and exciting religious vision.'

The American physicist and theologian Ian Barbour sees the modern scientific understanding of our origins as a useful corrective to wrong perspectives in the past.

'The sense of kinship with other creatures is very important. Maybe the Christian tradition has drawn too sharp a line between humanity and all other creatures,' he suggested. 'I think we've tended to see ourselves as *so* distinctive that the value of other creatures in the wider scheme, and in God's plan, is left out. So it's the sense of commonality I want to recover first. The ecologists teach us the importance of the whole community of life, and I think much of the Bible does too. The whole created order is of concern to God. God is interested in other creatures for their own sake, and not just for our sake (for us to exploit and use). So that is something the Church can learn from science, and I really value that.

'But having said that we're part of a larger community of life, I would also say we're a very distinctive part. Ninety-nine per cent of our DNA is the same as that of one of the higher apes, chimpanzees or gorillas. And yet that last one per cent – boy, what a difference that makes! We have a distinctive kind of language use; we have a distinctive ability to look to the future, and to interact with each other at the personal level. We're moral; we have ideals and purposes, and the kinds of abstract thought that only the human being is capable of. So I do want to maintain that kind of distinctiveness.'

Philosopher Roger Trigg points out that the need for us to come to terms with our having an animal nature is not exactly a new idea.

'We are evolved animals,' he agreed, 'but I would hesitate to say that we're *just* evolved animals. I suppose that's another example of what some people call "nothing but-tery": we're nothing but this, a mass of meat,' he said, pointing to himself. 'Of course we have an animal nature. This has been part of religious understanding. In the New Testament, St Paul talks about our "lower nature". Part of it we can be proud of; part of it is perhaps something that has its darker side, and needs to be controlled.

'It's at that point, I think, that an understanding of what it is to be human begins to move away from simply our animal nature. In other words, the raw material of our nature is provided by our genes. There's no doubt that that gives us predispositions to behave in one way or another. But what *distinguishes* humans from animals is our rationality. The very existence of science, the existence of socio-biology (which looks at human social behaviour), emphasizes the importance of human reason. Chimpanzees don't have a theory of chimpanzee nature; we do have a theory of human nature. In other words, we have the raw materials, but *we can rise above them.*

'I live in Stratford-upon-Avon. If I walk down the main street there, it looks as if all of the shops and buildings were built in different periods. In fact, a lot of them have had more recent facades built (eighteenth or nineteenth century) but are still timber-framed behind. In other words, the structure is the same but they look very different. Human nature is a bit like that. There are two strands in human nature: the genetic, and then the social answer to the genetic.

'I would also add that we have freedom and reason. We're not just puppets being worked by society; we're not just puppets worked by our genes; we're not even just being worked by a combination of both. We can reason about it, and even withstand *both* influences.'

Genetics

Roger Trigg touched there on the possibility that we are influenced in our behaviour by our inherited genes. Most, if not all, animals exhibit behaviour that is genetically influenced. It's behaviour they have not, as individuals, had to learn from personal experience. It's part of the make-up they were born with. For example, newly born animals instinctively relate to their mother, they automatically avoid predators, kill for food, and so on. Like the genes that control our physical characteristics, those that influence behaviour patterns are encoded in the famous DNA molecule.

Now one of the striking things about us humans is that, as already mentioned, we share a large proportion (99 per cent) of our genes with our close relatives the chimpanzees. So what is that telling us about ourselves? If it's obvious that the behaviour of other animals is genetically influenced, won't the same be true of us as well? If so, what kind of behaviour are we talking about?

Arthur Peacocke used to work on DNA. There was a large model of a DNA molecule on the coffee table in front of us as we talked in his lounge. I asked him about these behaviour traits. Presumably they were fashioned by natural selection. Did that mean they were of a kind that had survival value?

'The genes wouldn't be there unless at some time in our history they had survival value,' he told me. 'There are patterns of behaviour, our aggression say, which can obviously be associated with aggressive behaviour in animals. Now if you're very aggressive, you might think this is a good way of surviving. But in the biological world it's actually much more subtle than that. A totally aggressive animal would not survive because it would kill off its food sources too quickly and therefore fail to survive into enough generations. It's a balance which enables populations to survive. It's simplistic to say that certain characteristics we don't like (such as aggression) are due to biology. Care of the young is equally due to biology. Social cohesion which protects the weak is also part of biology. So it is not quite as simple as saying that all our bad behaviour, as it were, is due to our biology and all our good behaviour is due to something we've added to it.

'Of course, none of this *determines* what we do. Most people, unless they are very ill, do have some freedom of choice within the range endowed them by their genetics.'

Roger Trigg is likewise convinced that a significant part of our behaviour is genetically influenced.

'A lot of the inherited behaviour traits are, of course, to do with either our own survival or our ability to produce fit descendants. Care for offspring is going to be built into us in a very real way. Fear of danger may be built into us. It's often thought that a fear of snakes, for instance, can have a genetic impulse. Certainly, people who like stroking snakes aren't as likely to survive as people who keep clear of them! Fear of running water is another one. I remember being almost mesmerized standing on the edge of Niagara Falls, watching the weight of water slipping over. Again, people who like jumping into Niagara Falls aren't going to have many descendants. So all of this does produce a certain human nature that is passed on genetically.'

A key feature of religion is morality. Is this moral sense some-thing which might have evolved? Is there survival value in being, say, altruistic? Altruism in favour of one's offspring, yes. We see that happening: a mother bird, for instance, might deliberately attract the attention of a predator towards herself and away from her young. That makes sense in evolutionary terms because she shares much of the same genetic material with her young, and from the *gene's* point of view it is more advantageous for it to pass on and survive in the next generation than to persist a little longer in the older one. What is being selected out here is the gene itself, a gene that makes the mother bird behave in that way. So she is not really being as 'noble and self-sacrificing' as one might at first think.

But in addition to the type of altruism whereby we care for close kin, can biology account for the way we sometimes care for people who are *not* related to us? Philosopher Roger Trigg again: 'Socio-biology has emphasized two mechanisms: one is care for relatives, and the other is being kind to others – *so that they will be kind to you*. An example of the second, amongst animals, is mutual groom-ing. Animals will groom each other, but they will not groom an animal that doesn't return the compliment. They learn from expe-rience – and there's a lot of that in human nature. I mean, a lot of "morality" reduces to prudence: "Honesty is the best policy"; "Let's invite them to dinner, because they'll invite us back."

'But in the end, that doesn't take us very far with morality. Religion, and particularly Christianity, emphasizes not just love of one's own family, but love of one's neighbour, where one's neigh-bour is thought to be *everyone*. One example is Mother Teresa helping people who can't help her back, and who certainly aren't relatives. She is a prime example of what I mean by "morality". Socio-biology can explain a lot of basic impulses, but it can't explain human morality – caring for others regardless of what one's getting out of it. There's a very definite question there about the source of that morality.'

'What would you say then was the ground for this other dimen-sion of morality?'

'I think that it begins to point to what religious people have traditionally talked of as "soul" or "spirit". I think there is a dimension of the human personality which goes far beyond anything that other animals share in, and indeed goes beyond what can be investigated by science. Science is itself the product of human reason. It can't, in the end, explain human reason away without explaining itself away. Human beings somehow have the ability to recognize truth. They have the ability to recognize it in the physical sphere (in the area that science investigates) and also in the moral sphere. Once one begins to question that, I believe one is actually questioning the ability of science to say anything true *itself*.'

Will Provine has an unconventional view of human nature. He sees our behaviour as not so much influenced by genetic factors as determined by them.

'Humans are comprised only of heredity and environment, both of which are deterministic,' he asserted. 'There is simply no room for the traditional concepts of human free will. That is, humans *do* make decisions and they go through decision-making processes, but all of these are deterministic. So from my perspective as a naturalist, there's not even a possibility that human beings have free will.'

This greatly puzzled me. Did it mean he went through life never making a decision?

'Not in the least,' he assured me. 'Even computers make decisions. They can be programmed to play chess, and then the computer must make very many decisions. We humans must make decisions as well. But simply because I make decisions does not mean that I have human free will.'

'So are you saying that when we *think* we have two possibilities open to us, this is actually an illusion – there is in fact only one choice of action?'

'The choice of action is determined only at the moment of choice. So if you think in terms of a tennis game, the game is not all determined before it begins. The player makes the decision at the last split second as to where to hit the ball. And all of our other decisions are similar to that. They're only determined at the moment when the decision is made. So no, I do not see us living in a completely deterministic Universe in which everything was determined from the beginning. Rather, I see humans as being fairly *local* objects that are determined. Local determinism, in other words, rather than some kind of universal determinism.'

I didn't really understand the distinction he was making between local and universal determinism. Nevertheless, I had to concede that there is a very real problem about understanding where our sense of free will comes from. Quite apart from arguments based on heredity and environment, if I as a physicist regard the human body as made up of matter, and that matter is subject to the same laws of nature as inanimate matter, then it is difficult to know where free will gets a look in – why our actions are not wholly determined by the remorseless working-out of those laws. And yet, from the point of view of regarding myself, not as a physicist now, but as a conscious human being, it would be absurd for me to accept such a conclusion when it comes to how I live out my life.

I continued, 'You're saying that our decisions derive either from heredity (which is no responsibility of ours) or from the environment (which again might not be any responsibility of ours). Are you saying then that nobody is ever to be held responsible for what they've decided to do?'

'That is correct,' he declared. 'There is no ultimate responsibility that can be laid at the feet of human beings. One can be conditioned or trained or educated to be morally responsible, and I believe that moral responsibility is essential to enable us to live together in society. But that's *programmed* into us by others around us, by cultural traditions, and so on. There's no hint to my mind that the moral responsibility exhibited by human beings comes from any freedom of the will. I think that free will is not only unnecessary but very destructive as well. We use it as an excuse to punish terribly those who have done things wrong in society. When someone does something wrong, it appears to me that the person is poorly programmed and needs to be reprogrammed.'

'So are you saying that, for example, a prison sentence should be regarded as a reprogramming?'

'Definitely. Prison sentences should be reprogramming, and

when an individual has been reprogrammed and can be a productive and contributing member of society, then I believe there's no justification for keeping that person prisoner any longer.'

'A corollary of what you're saying seems to be that one should never feel guilty because it's never one's own fault – and the obverse is that one should never take a pride in anything one has done because, again, you can't really take the credit for anything.'

'Well, that's not entirely true, because we can raise people to feel guilty after they've done certain acts that are antisocial. It seems to me that we *can* have a sense of moral responsibility, and have that connected with feelings of guilt. But it's all programmed. As for feelings of accomplishment when we do something wonderful, I do indeed believe that we tend to give ourselves more credit than we deserve. When we think about our own individual accomplishments, we need to look more carefully at the combination of hereditary and environmental advantages we have had.'

After this, it was a relief to find someone who, like me, profoundly disagreed with Provine's thesis: John Habgood.

'I think it's clear that our general characteristics as individuals do depend, to some extent, on our genetic inheritance. But that's a long way from saying that we are genetically determined. We're subject to a great many influences. One of them is genetic; one is environmental; and one is what we make of ourselves. And perhaps the most important is the element of free choice – of taking responsibility for ourselves. This is the heart of morality. With animal behaviour, it is more difficult to talk about deliberate intentions. The essence of human behaviour is that it is intended; it is thought about. And that's the level on which our moral responsibility becomes so significant.'

'If, to some extent, we are genetically influenced in our behaviour, what do you reckon that kind of behaviour would be? If it was part of the survival kit of our ancestors, what kind of behaviour would this give rise to?'

'Some interesting work has been done by socio-biologists trying to study the evolution of behaviour. They've pointed out that there might be a certain natural tendency towards co-operativeness, particularly in the higher animals. There's a lot of discussion about whether there is an innate aggressiveness in human beings. Indeed, there needs to be a certain aggressiveness if animals are to survive.

'The point is, what do we do with the capacities we have? For example, if you didn't have any aggressive people you wouldn't have any successful businessmen. That might be a good thing or a bad thing, but aggressiveness certainly underlies a lot of what they do.'

'You mentioned co-operativeness (or altruism, as it's often called). I can understand how altruism could have evolved between you and a close relative of yours, someone with whom you share much of the same genetic material. But can there really be altruism amongst individuals that are *not* closely connected genetically?'

'No, that's a difficulty. But even if you accept that there may be a genetic basis for altruism – for example, blackbirds which give warning calls to others – the blackbird isn't actually intending to do that. It's not acting morally. It is simply responding. The response happens to be helpful to other blackbirds around, telling them that there's an owl or something nearby. One has to be very careful not to slide too quickly from this kind of genetically based "altruism" to what in human beings we call altruism – that being a matter of deliberate choice.'

Arthur Peacocke approaches altruism in the following way. 'Altruism, in the Christian sense, means that a person lays down his life for others *regardless* of their genetic connections. It could be argued that the religions of the world have been those systems of thought which have transformed biological tendencies (to protect only our genetic kin) into our wider sympathies and commitment to the life of human beings *in general*. The work of individual persons is then valued not because of their genetic connections, but just because they are human beings. There are those who argue that it was the role of the great religions of the world to provide the framework of altruism in that broadest moral sense. I think that "altruism" is a very slippery word. It has a technical meaning in biology, and must not be confused with this wider meaning in the moral world.'

'If one thinks of altruism in this wider sense, can you see any survival value in the creation of a gene which promotes that kind of behaviour?'

'The "you scratch my back and I'll scratch your back" kind of gene? That is one of the explanations socio-biologists bring to bear on certain mutually reinforcing behaviours they observe. And I think it's a good explanation within its own limited context. But the real question is whether this a *sufficient* explanation for the complexes of relationships human beings get involved in. I think it's still a very open question. I don't think it's been shown at all that socio-biology can explain the wider moral aspirations of humanity.'

'Do you see religion functioning just as something that is *useful*, something that promotes desirable, altruistic behaviour? Or is it ultimately theistic – rooted in the existence of a God?'

'I would be the last to want to justify religion simply on functional grounds: a useful fiction to enable people to behave appropriately. There are some biologists who are beginning to argue in such terms: that we believe in ethics and morals simply because it's a useful fiction. But I don't think that's adequate. I'm a theist because I do think there is an all-pervading and all-surrounding Reality (with a capital R) which is other than the world we observe.'

Future evolution

So far we have considered how we humans got to our current stage of evolution, and how much of our present make-up we owe to what happened in the past. But evolution is an ongoing process. It is happening all around us. We can observe how certain other species have undergone noticeable changes over the period they have been studied. So what of the future? Are humans still evolving? Shall we eventually develop into a superhuman race?

Richard Dawkins has his doubts: 'Evolution is such a slow

process. Even if humans are evolving towards being superhuman, we have to remember that the time-scale on which that would happen is going to be far, far longer than anything that can be of interest to us with our historical perspectives. So if you want to see a superhuman race (with big brains?), it will not be in a few centuries, it will be in a few *million* years. It's almost like saying we're static. In effect, we're like the hour-hand of a clock that you can't see moving.

'I don't know whether humans are still evolving. In one rather trivial sense, ordinary natural selection is at present being frustrated by the fact that it's jolly difficult to die young. Nearly all of us live long enough to reproduce. Whether or not we actually reproduce depends not upon whether we're good at surviving, but upon wholly extraneous conditions having to do with culture – considerations that have nothing to do with the ordinary Darwinian world. So if you ask why some people have a lot of children and others not, it will be for reasons which have no very great evolutionary interest – religion, or culture, or education. And therefore, if there are selection pressures going on in modern Western society, they will be pressures of an odd kind.'

'If one were to characterize the theory of evolution as "survival of the fittest", do you see any problem over the fact that we take such great care of people who are sick and dying? Are we storing up problems for ourselves (as an evolving animal) by taking such care over what one might call "the unfit"?'

'That's an aspect of what I've just been saying: that it's hard to die young. And those people who, as it were, do their best to die tend to get saved by doctors. So there is a kind of thwarting of natural selection, as you say. There will be a tendency for medical science to preserve in the gene pool genes which would otherwise have been eliminated.

'I prefer not to be worried about that, because I think I'd rather live in a world of doctors than in a world without doctors, and I'd rather people were saved than not. So I think it's a price we have to pay for the benefits of civilization, and I think it's worth it.'

*

As Richard Dawkins pointed out, evolution by *natural* selection is slow. But is there not a way of speeding things up? There is; it is called the Human Genome Project. I asked Ted Peters, Professor of Systematic Theology at the Lutheran Seminary in Berkeley, California, about it.

'The Human Genome Project is a spectacular adventure in research science that rivals the moon-shot of the 1960s,' he explained. 'It's a worldwide, co-operative effort by molecular biologists and other geneticists attempting to sequence the DNA in our body cells. They want to sequence and locate all of the genes located there, maybe 100,000 of them. The purpose of this, of course, is to get new knowledge, to climb the Mount Everest of scientific mystery.

'But the real pay-off for society is going to be tremendous advances in medical science and medical care. Many diseases are due to genetic predispositions. If we can locate the gene for these particular diseases and come up with a pharmaceutical to keep them turned off – or to turn them on at will – the advances in human health will be just staggering.'

'I don't think anybody would take issue with helping people to overcome diseases,' I said. 'But doesn't the same project hold open the possibility of giving us even greater power than that? Doesn't it mean that we could start to engineer genetically the kind of people we want: intelligent people, non-aggressive people, that sort of thing?'

'Currently, that's science fiction. But we *are* getting closer and closer all the time to making fiction into reality. If we have genes for diseases, we also have genes for human characteristics, probably eye colour, hair colour, intelligence, height, strength, all that kind of thing. And the next question is, "Are there genes that govern human behaviour, such as aggression or sexual proclivity?" Certainly there's a predisposition to alcoholism. And in 1994 a gene for aggression among men was found. Where might that lead? That's a good question.'

'Is anybody taking a lead in deciding what kinds of engineering of the genes are acceptable? I suppose *somebody* has to make a decision – for example, as to whether we are going to make people who are not aggressive. Who makes that decision? And would it, in fact, be a good thing?'

'This is sometimes called the "Brave New World scenario". In Aldous Huxley's *Brave New World* there was a totalitarian government that made these kinds of decision. That will not happen in the near future, because currently our genetic science is market-driven. The biggest investment is being made privately, through pharmaceutical companies and other kinds of capital venture. I think what's going to happen is that services in genetic manipulation will simply be sold on the open market, especially through clinics engaged in reproductive technology. So there will not be any uniform national or international programme. In fact, I see that as a bit of a problem. It will be people's taste and preference, and what products can be sold for a profit, that will determine exactly how we engineer people in the future.'

'I can well imagine a situation arising in which a government considers the cost of crime to society to be so high that it would be happy to put a few dollars into genetically engineering its population so that it didn't have criminal tendencies. Would that be a possibility?'

'That is a very real possibility. And the way it's most likely to happen is going be less than genuinely scientific, in my judgement. Society will jump to conclusions and probably reinforce existing social injustices, especially racial prejudice. If, for example, there's a gene for urban crime, and we then notice that there's a disproportionately large number of black people in prison, the assumption might well be – without doing any blood tests – "If you are black, you have the gene for crime. Therefore we should incarcerate you in order to protect the rest of society." You will save a lot of government money by just incarcerating a large number of people, and end up reinforcing an existing unjust social structure. That's the kind of deleterious thing that could happen.'

'As somebody who has been involved in this sort of thinking for a long time, how would you react to the possibility of genetically engineering people who were non-aggressive, non-selfish – indeed, free of sin (to talk in a religious sense)? Would such a goal be desirable? Or would you say that evil and suffering are key ingredients of what it is to be human?'

'Well certainly, evil and suffering are key ingredients of living human life as we currently experience it,' he replied. 'But I wouldn't put their removal on a list of desirables and deliberately programme evil and suffering out of our lives. The struggle to overcome them is just as human as to be victimized by them.'

I reminded Richard Dawkins that he had earlier said evolution through natural selection was a slow process. But, I asked him, could we not now, through the Human Genome Project, accelerate matters by going in for *directed* evolution?

'The Human Genome Project could be thought of as both a pure research activity (where you're finding out what the sequence of DNA in humans is) and as having an interventionist aspect: on the basis of that knowledge, one might interfere and direct evolution, as you've said. Directing evolution has, in a way, always been done, but not by directing the mutational side of it. It's been done by directing the *selection* side. Artificial selection has been practised in agriculture, dog-breeding and so on for a long time. Not by genetic engineering in the sense of inducing mutations, but in the sense of waiting for the random genetic variation to come along and choosing some variations rather than others for breeding. Adding the ingredient of directed mutation is going to speed things up a bit, but it's not going to be *qualitatively* different from what we are already familiar with when we breed poodles and Pekineses and cabbages and things. For a long time it's been theoretically possible for us to do artificial selection. Instead of breeding Pekineses, we could have been breeding humans. And slightly oddly, we've never actually done this. If, say, a country wanted to win the high jump at the Olympic Games in two hundred years' time, there is an easy way to do it. You've only got to mate expert male high-jumpers with expert female high-jumpers. And you could do the same for mathematicians, I conjecture, and for musicians. But it's never been done. There is widespread hostility to the idea.'

'Where do you think that hostility comes from?'

'I think it comes largely from the Hitler experience. Hitler was keen on the idea, and so it's had an understandably bad press ever since.'

Another biologist, Steven Rose, is disturbed by some of the claims being made on behalf of the Human Genome Project.

'I think that the reductionist way of viewing the world is very clearly indicated by the tremendous enthusiasm amongst geneticists and molecular biologists for the project. It's very instructive to look at the metaphors they use to talk about it: they refer to DNA as the "code of codes", the "book of life" – they even use an almost religious metaphor, "the Holy Grail".

'A fundamentalist, reductionist metaphor runs right the way through. I find this very troubling indeed. Essentially, living organisms (but particularly humans) are multi-dimensional. DNA is a one-dimensional strand of letters; humans are at least four-dimensional – three dimensions of space and one dimension of historical time in which we develop. You can't easily extrapolate from one dimension into three or four dimensions. As biologists, we do not know the rules by which you translate the single dimension of DNA into the three or four dimensions that make up any living organism. And until we know those rules, to assume that we can understand all of humanity (whether you are likely to be violent, to be religious, to be politically radical or conservative, to be homosexual or heterosexual) by doing DNA sequencing is naïve in the extreme. When it's not merely naïve, I think it also generates a dangerous ideology. Exciting as it is scientifically, I find the claims of the Human Genome Project really quite dangerous.'

I next asked Roger Trigg whether he had any worries, or excitements, about the Human Genome Project.

'It's of immense interest and profound significance,' he said. 'In the immediate future, its main interest lies perhaps in its ability to identify the genetic causes of disease. In the long term, it poses greater problems because it gives us the ability to interfere with human nature. And, of course, the question is, who is doing the interfering? Who's going to be looking after the scientists doing the genetic engineering? So I view the Genome Project both with interest (the idea that it will help human beings) and with some alarm as to the far future because of the powers it might give us.'

I asked him whether it would be a good thing genetically to engineer a population that had no criminal tendencies. It was not a prospect he welcomed.

'It sounds exceedingly frightening. One is putting oneself in the position of making decisions about what people *ought* to be – a bit like playing God. Unless we're omnipotent and really in God's position, and know exactly what we're doing, then who knows what we shall achieve? One of the problems about genetic engineering isn't so much what you mean to do; it's what you in fact do but didn't mean to.'

Throughout these discussions it was taken for granted that the human species will indeed continue to evolve, whether through natural or directed evolution. But is that necessarily the case? I put it to Steven Rose that one of the things about species is that they tend to die out. Did he see the human species as having a finite life, and should we as humans be taking this into account?

'I don't have a great sense (as some writers do) that humanity can't be allowed to die out – that it's got to go on and colonize other planets, or whatever. All evolutionary history,' he said, 'tells us that species live and die. They die out as circumstances change. Even if we knew that a meteorite was going to crash on the planet a century from now, we couldn't change our biological make-up in such a way as to enable us to survive beyond that. So yes, the human species will die out. I will die as a human, and that mortality is an integral part of being a living organism. And I hope I face that with a fair degree of equanimity. Certainly, as I get older I recognize more and more bits falling off my body – bits falling off my brain – and there is a certain unpleasantness about it. But I see it as a biological inevitability, and I don't think it troubles me. Whether it will trouble me when I am not in my upper fifties, but get to my upper sixties, my upper seventies, and my upper eighties, I don't know!'

'I suppose it's all part and parcel of the evolutionary process that we have to have death and replacement,' I said.

'Oh, absolutely. I mean, there are very good biological reasons why organisms die.'

One good reason why there might not be any significant further human evolution is that our particular species has evolved to the stage where we have language; we can share experience and build on the accumulated knowledge of past generations. This process has culminated in the discovery of nuclear power. Though we might at present be going through a relatively peaceful period in the history of humankind, can we be sure, given our past, that there will not be at some point in the not-so-distant future (as reckoned in evolutionary terms) an all-out conflict in which one side or the other does not shrink from using nuclear weapons? Perhaps our destiny is not that we as a species will develop into superhumans, or that we will die out in some natural way, but that we shall one day, in a moment of madness, annihilate ourselves. Indeed, this might well be a universal scenario constantly re-enacted throughout the cosmos, wherever intelligent life (so-called) develops. We just get too clever for our own good.

Extra-terrestrial life

That last comment raises the question of whether life has indeed evolved anywhere else in the cosmos. I tackled Rod Davies of Jodrell Bank about the possibilities. 'When you consider the enormous number of stars out there, each of them being a sun,' I suggested, 'many of them presumably have planets going around them?'

'There are several that have pretty clearly identified planets going around them,' he replied. 'And there's no reason to think there aren't *a lot* of these.'

'So what would you say are the chances of there being extra-terrestrial life?'

'I would have thought rather high,' he ventured. 'The building materials for life are distributed right through our own galaxy and many other galaxies. Whether we'll ever have contact with other civilizations elsewhere in the Universe is a very interesting question. People are actually setting up experiments to see whether they can get signals from other intelligent civilizations around the universe. They haven't succeeded as yet, but it's a possibility worth following up. It may be that some civilizations have peaked, and others have not yet got to a technological level where they can communicate with us. When you think of it,' he said, glancing out of his office window at the big radio telescope, 'we've only been in a position to listen with radio, or use sensitive optical systems, over the last fifty or a hundred years at the most. So there's a very small time-slot in the fifteen billion years total for an overlap between a civilization trying to contact another civilization. But I think the chances are very much that there are other planets with life on them, in some form.'

'Does that in itself affect the way you see human beings and our own importance?' I asked.

'From a religious point of view, if you are looking at the way in which God created the Universe, I don't see there's any reason why he (or she) should not have created a universe in which there were multiple births of life in one form or another. I think that would be very much a part of the pattern.'

How did Richard Dawkins view the possibilities of the same process of evolution by natural selection giving rise to intelligent life elsewhere in the cosmos?

'I find that a very intriguing question,' he replied. 'One's first thought is that all the lessons of history teach us that we should not be parochial, and we should not think there's anything special about us. We used to think we were the centre of the Universe and now we know we're not. By the same token, just because we're the only form of life we know about, we should be very, very wary of suggesting that there isn't any other life. One argument which I put in my book *The Blind Watch-Maker* – that there may actually be only one form of life – is quite an interesting point. If the coming together of atoms into a self-replicating molecule is a very, very, very rare event (which it could be), such that it's only ever happened once in the Universe, then it's not surprising that we're the place where it's happened. We've *got* to be the place where it happened because we're the people talking about it. You can't just say that, because it happened to us, it's probable. You have to look at the event itself and ask, "Is it an improbable event?" That's a job for a chemist. Those chemists who have thought about it tend to say it's fairly improbable. But when you match that up against the number of planets in the Universe where it could have happened, the improbability pales into insignificance beside the sheer number of planets. So probably there is life elsewhere.

'As to whether it's intelligent, you can use a similar kind of argument. You need a nervous system. But even a nervous system isn't enough. Perhaps you need a particular kind of nervous system. Perhaps you need something equivalent to language to get culture going. It comes to guesswork. And for what it's worth, my guess is yes, there probably is intelligent life elsewhere in the Universe. But it's probably so well spaced-out that we're unlikely ever to meet it.'

Paul Davies regards the odds on there being life elsewhere as high, and thinks its discovery would be very significant.

'It's inevitable that if we discover life elsewhere in the Universe, it will change for ever our perspective of our own species and our own planet. I think the discovery of even a humble bacterium somewhere else in the Universe (if we could be sure it had evolved independently of life on Earth) would be momentous. Those people who cling to the idea that humanity is the pinnacle of creation, or that somehow we were made in the image of God, would I think receive a rude shock.

'If we make contact with an alien community, for example by picking up radio signals, it's overwhelmingly probable that they're going to be very much in advance of us technologically,' he continued. 'This is simple statistics. The probability of two civilizations reaching the same level of development after billions of years of evolution are just infinitesimal. So other communities are either going to be a long way ahead of us or a long way behind us. We won't be able to detect the ones who are a long way behind, so we're almost certainly going to be dealing with civilizations that might have been around for millions, or even billions of years.

'It seems clear that not only would these beings be exceedingly far advanced of us technologically, but also, in the broadest sense, *spiritually*. So it would be a very humbling, but in some ways inspiring experience to be in contact with them.'

'Why did you say we would have to revise our understanding of ourselves as being made in the image of God?' I asked. 'After all, when we talk about this, we don't mean that we *physically* look like him.'

'Well, some people believe that, of course.'

'I know. But if one interprets it along the lines that we have some of the qualities of God, why can't aliens which look very different from us also share those qualities?'

'Well, they can ...'

'If they're in contact with the same God?'

'Yes. Certainly we can conceive that they do. But this does lead us into all sorts of difficulties. Christianity, in particular, has difficulties with regard to the very special role that Jesus Christ plays. The Church should give a lot of attention to exactly how it wishes to regard possible alien beings. If they wish to retain Jesus Christ as the saviour, is he the saviour of mankind only, or of all sentient beings throughout the Universe? Or will each community have its own saviour? Doesn't it all start to become a little bit ludicrous?'

'No. I wouldn't say that,' I objected. 'What Christians believe is that the Son of God became a human being. I can't see why the same Son of God, who has existed for all time, can't take on the form of other creatures once they reach the stage where they can communicate with God.'

'I find it mildly absurd that this drama of the life and death of Jesus Christ is re-enacted planet by planet throughout the cosmos. It just seems to me not a very satisfactory resolution of the problem.'

We clearly weren't going to see eye-to-eye on that one, so I took him up on another point. 'Earlier you said that aliens could be advanced in various ways intellectually, and you also mentioned that they might be *spiritually* more advanced. What do you have in mind there? As far as I'm concerned, spirituality is a measure of how close you are to God, and how loyal you are to him. As human beings, we've already reached the stage where someone like Christ is prepared to lay down his life for God. I don't see how you're going to improve on that.'

'But there was only one Jesus Christ,' he countered.

'Yes, but many saints have been martyred. They have also laid down their life for God.'

'We're still talking about exceptional human beings, though, aren't we? If you were to look around today, you would regard a lot of people as severely lacking in a spiritual dimension.'

'Certainly,' I agreed.

'So what I would hope and expect is that as life develops and evolves (if it doesn't destroy itself), both in the conduct of their lives and in their relationship to nature and to God, these beings would reach a level of development well ahead of what we see on Earth today. But you're quite right: individual people seem to be pretty far along that road.'

I invited him to comment on the possibility of nuclear holocausts occurring at some stage soon after each alien species had reached an advanced level.

'The people who search for extra-terrestrial life certainly raise this as a hypothesis,' he said. 'It may be that the duration of a technological community is limited to a few decades. In which case, even the optimist would admit that at this particular time we're probably alone in the galaxy – alone in the sense of being the only advanced technological community.

'I'm an optimist. I think there's no reason why we can't come through this rather nasty phase of our existence, when we have the power to destroy ourselves but maybe lack the political or social organization to prevent such a catastrophe.'

*

Ted Peters also thinks that the existence of life elsewhere is plausible. Did he think this raised any issues for religious belief?

'There are some in the scientific community who keep making the absolutely ridiculous claim that, if we were to discover intelligent life on other worlds, religion – especially the Christian religion – would be destroyed. I've been curious as to why these stupid things are being said. They say the Christian religion teaches that human beings are the centre of the world. Christianity has never taught that. There is nothing within Christian theology, or Jewish theology, that would find extra-terrestrial life at all surprising.'

Ted being an expert in both genetics and extra-terrestrial life (as well as being a theologian), I asked him whether he thought alien civilizations – in common, perhaps, with ourselves – were likely to be driven to nuclear destruction by an aggressive genetic make-up.

'Human aggressiveness has to do with anxiety. We feel that our existence is being threatened and we strike out in violence. Even when we are *not* threatened, we imagine that we are and engage in violent behaviour. What about extra-terrestrial intelligent beings? Are they likely to have anxiety in the same way that we experience it? It is very likely that they would.

'Could it be that they are more *virtuous* than us? Maybe they have found a richer philosophy, or have found some kind of social way (or a stronger police force!) to get better control of themselves than we in the human race have. But this is really speculation.'

'If there is such a thing as extra-terrestrial beings, do you expect them to show signs of being religious?'

'Yes, I do,' he replied. 'Religion is a form of culture in which we try to understand the meaningfulness of life in its largest possible scope. So I think they are likely to be religious, in the very generic sense that we are "religious". From the point of view of a Christian theologian, would they be the beneficiaries of a special revelation from God in one fashion or another? Again, I would just have to wait and ask them.'

I thought I would give the last word on extra-terrestrial life to John Habgood. Did he think it was likely?

'Nobody knows,' he said. 'Within my lifetime there have been wild speculations both ways. Astronomers have confidently come up at one point saying there must be millions and millions of planets like ours. Then ten years later they're saying something quite different – that we are unique. Theologically, I don't think it matters.'

'You don't feel it affects the status of human beings to think that there might be teeming masses of life-forms out there?'

'Not in the least. Surely God is large enough to cope with a variety of creatures. Indeed, there was a fascinating poem written at the end of the nineteenth century, when this sort of speculation first came up. It was by a Mrs Hamilton-King (who I think is not known for anything except this poem), and in it she tackled the question of life on other planets, producing the phrase: "God may have other words for other worlds, but his word for this world is Christ".'

'So,' I asked, 'if there were living forms out there as advanced as ourselves, they would have the Second Person of the Trinity, the Son of God, visiting them too, as one of them?'

He looked dubious. 'I think one needs to put it in more general theological terms. Clearly, if God is going to relate to other creatures in other parts of the world, this would be through the Second Person of the Trinity. Therefore, the Second Person of the Trinity is, as it were, God expressing himself in relation to the world that he has created. For us, that expression is in terms of a human life, the life of Jesus Christ. For other worlds it could be different. We don't know.'

'Suppose there were life-forms somewhere else in the Universe which were, in evolutionary terms, far more advanced than ourselves. How would that affect human beings in the eyes of God, do you think?'

He smiled. 'A father may have clever children and stupid children,' he said, 'but he loves them all just the same.'

Chapter 3

The Mind

Some of the severest challenges to religious belief, and to our understanding of the status of human beings, have come from the field of psychology, and from related investigations by neuroscientists into the workings of our brain.

Sigmund Freud

No one has had a greater impact on psychology than Sigmund Freud. And as is well known, he was of the view that he could explain away religion.

My journey into the fascinating world of the human mind began with psychologist Anthony Storr at his home in Oxford. I started by asking him to paint a picture of Freud the man.

'Freud was an extremely conscientious man,' he told me. 'He was incredibly clever and hard-working. He was very successful at school. He became an accomplished linguist in a great many languages. He was very well-read. He knew quite a lot about sculpture, a little about painting. He collected statuettes. He hated music. He was superstitious – about numbers. He was a compulsive worker. He saw eight or nine patients a day. He would then stay up and do his writing after supper. He went to bed about 2 a.m.; he did that night after night. He had very little relaxation, apart from playing chess. He did take decent vacations, and then he would go for long walks in the mountains and that kind of thing.

'He was an extremely fluent speaker. He could speak without notes for two or three hours at a stretch, and hold an audience in a trance. Many found him rather formidable. But most people were convinced that he was a very honest man, rather pessimistic, hard-headed.'

'Was he a touch arrogant?' I asked.

'I don't think I would call him "arrogant". He was very sure of his own abilities, but then he did have very considerable abilities. He was certainly an extremely persuasive writer. I mean, you can hardly read a good paper of Freud's without being convinced – temporarily.'

'How did Freud try to explain religion?'

'He repudiated religion altogether. He wrote a book called *Future of an Illusion* about religion. He believed that it was a wish-fulfilment, an escape from reality, something which men invented because they couldn't stand the terrors of life. He realized, of course, that virtually all cultures have had religious beliefs of some kind. But he thought their function was to control fear.'

Anthony took down a book from the shelves and began reading: 'He says here, "The Gods have a threefold task: they must exorcize the terrors of nature; they must reconcile man to the cruelty of fate, particularly as is shown in death; and they must compensate them for the sufferings which a civilized life in common has imposed on them". So all those functions of religion are why Freud thought man had such beliefs. He thought that religion really originated in people's feelings of helplessness.

'As an adult, man is confronted with all kinds of things which threaten him, from earthquakes to disease. As a small child, he is even more vulnerable and helpless. Freud thought that "God" was a substitute for the father. He said (and I'm quoting him), "The derivation of religious needs from the infant's helplessness and longing for the father, seems to be incontrovertible – especially since the feeling is not simply prolonged from childhood days, but is permanently sustained by fear of the superior power of fate. I cannot think of any need in childhood as strong as the need for a father's protection." So that was his explanation of why people had religious beliefs: they were looking for an imaginary father in Heaven who would protect them and look after them.'

'Freud laid great stress upon the early years of life, and how the events that occurred then can have later repercussions,' I said. 'Were there any events in Freud's own early life which might have coloured his attitude towards religion?'

'Only the fact that he was brought up in an irreligious household. I doubt whether they observed any of the customary Jewish festivals. Although he very firmly hung on to his Jewish identity, and belonged to Jewish organizations, and was proud of being a Jew, his actual religious *practice* was practically nil. So that's why I think in some ways he never got religion in perspective.'

'I understand that he had a Catholic nanny, and that when he was aged two-and-a-half this nanny was caught stealing from the household and dismissed. Do you think that could have made some kind of impression on Freud's mind concerning what these so-called religious people get up to?' I suggested.

'I don't think the incident over that nurse influenced him very much – it was rather interesting, though. I mean, his whole concept of religion was based on the *father*. He didn't have any general religious appreciation. He didn't appreciate, for instance, that Buddhism doesn't require belief in God or gods. It was a blind spot.

'And I think another reason why he couldn't appreciate religion was that he was quite incapable of having – or *said* he was incapable of having – anything approaching mystical or ecstatic experiences, feelings of unity with the Universe as a whole. Many people have interpreted such experiences in religious terms; they've thought that God was directly speaking to them, that it was a *religious* experience. But Freud interpreted it as a regression to a very early state in which the infant can't really distinguish itself from the mother. The world is one in which the infant and the mother are not differentiated. It's also very like being in love, in which you feel in complete rapport with your beloved. Freud thought that *that* was all illusory. Being in love is a kind of insanity, and religion is a kind of illusion.'

I asked whether Freud regarded his psychology as a science. Anthony replied that Freud himself certainly thought that he was a scientist: 'He said something like "Our science is no illusion, but an illusion it would be to suppose that what science cannot give us, we can get elsewhere". He was very well trained in science. He knew what experimental science was like. He really wanted to go on being a researcher (he was studying the nervous system of dogfish and that sort of thing), but there was no money in being a research assistant. He had to give that up and go into medical practice in order to get enough money to marry. But he maintained that what he did in psychoanalysis was scientific.

'I myself don't think he should have claimed to be a scientist. There is absolutely no reason to suppose that such a claim can be sustained, and I don't think anybody now believes it. I mean, his conclusions can't be repeated; you can't predict from what he says. It doesn't fulfil any of the criteria of science.

'If psychoanalysis is not scientific, what *is* it?' he continued. 'It's a way of making sense out of things which would otherwise appear quite irrational (like neurotic symptoms). It's a way of making sense out of a certain variety of human experience. I think that's much the best way of looking at it.

'The other thing,' he added, 'is that if you're going to help any-body in psychotherapy, you've got to be able to make a relation-ship with them. So in a way it's very much concerned with the study of interpersonal relationships, as shown in the relationship between the analyst and his patient. For instance, if somebody has always been scared stiff of other people, or has had a brutal father who knocked them about, they come to a new person with certain expectations. They expect that they will be treated badly or harsh-ly, or not regarded. One of the learning experiences that you have in psychotherapy is to discover that there are people in the world who *can* accept you, like you, help you, and care for you, however badly you've been treated and whatever awful things have hap-pened to you in the past. That's one aspect which is perhaps not sufficiently emphasized: it is a study of interpersonal relationships.'

This conversation intrigued me. I wanted to learn more of Freud. So I decided to visit the house in Hampstead, London, where he spent the last two years of his life. There, in Freud's study, I met up with Montague Barker, consultant psychiatrist and Medical

Director of Heath House Priory Hospital, Bristol. It is a wonderfully atmospheric room, with its books and rugs. But it was the sculptures that caught my eye. There were so many of them, rank upon rank covering every square inch of shelf space in the display cabinets.

'They are absolutely magnificent,' Monty commented. 'But as a psychiatrist coming in and looking at this – with them all meticulously arrayed in rows (even the objects at the side are specially arranged so that they are exactly replicated on both sides) – then all I can say is there is a remarkable obsessionality about the man!'

There were even three rows of sculptures on Freud's desk, leaving rather little space for him to work. Also on the desk, placed on the blotting pad, were his wire-rimmed spectacles. They looked for all the world as though they had just been casually left behind while their owner had slipped out of the room for a minute.

But, of course, the real focus of attention in the room is the famous couch, brought all the way from Vienna. To one side of it, behind where the patient's head would have been, was the armchair in which Freud himself sat as he listened to what his clients had to say about themselves.

I experienced an irresistible urge to sit on the couch. I confided this to the curator, Erica Davies, who assured me that this was quite common. She told me how visitors were expected to stay behind the ropes near the door – they were not allowed too far into the room. But when there happens not to be a security guard present, some visitors are observed (on the closed-circuit television) to look around furtively, hop over the rope, rush over to the couch, sit on it for a second, before sheepishly returning.

'We call them "jumpers",' she said with a laugh.

Admitting to her that I was myself a potential jumper, I asked whether there was any chance of Monty and I conducting the interview sitting on the couch, seeing that it was for the BBC. To my delight, she agreed to make an exception.

I began by asking Monty what he thought was the legacy that Freud had left us. Did we actually understand ourselves better as a result of Freud's work?

'He highlighted something which was known, but never elaborated in quite the same way as he did: the way in which what we think is influenced by factors which we don't reflect on – what he called our unconscious. Also, there was the way in which we can project so many of our own unconscious desires on to situations and objects. That is very definitely a legacy from him.'

'And what relevance do you think this has for religion?'

'I don't think there is any doubt that when he said that we projected our ideal father on to some being in the sky, he was right. There are many people who *do* create a god in their own image, a father-god in the sky. Men have always created gods in their own image.'

'What was Freud's attitude towards religious experience? How did he explain that?'

'His view, quite simply, was that it was nothing but a psychological projection of our own wishes externally. He used the phrase "nothing but" several times when he was talking about religion. And that of course is the problem. Once you say "nothing but", you are not really being scientific any longer. That is part of the difficulty we've had over Freud's preoccupation with religion: here, the scientist in him seemed to be taken over by a "nothing buttery" approach.'

Fraser Watts is lecturer in Science and Religion at Cambridge University. His background is as a psychologist. I asked him whether Freud's own atheism was something which came out of his psychoanalytic studies, or whether it predated them.

'As far as we can see, it predates them. He was using his psychoanalytic ideas and framework to put a rational gloss on what I think was a gut atheism that came out of his personal background. In some of his more open-ended moments, he admitted that his psychoanalytic views could have been used to develop a completely different approach to religion from the one he himself developed.'

Malcolm Jeeves, a psychologist from St Andrew's University, expanded for me on Freud's idea of religion as wish-fulfilment.

'Freud asked himself the question: in the normal development of a child who has been growing up depending so much upon his human father, what happens when he gets to the age of sixteen – or whenever adolescence is these days – and feels he's got to assert his independence from his human, earthly father? He's got a conflict situation. He values his dependence on his earthly father, but at the same time, if he's to be an individual in his own right, he's got to assert his *in*dependence of him. So, said Freud, he solved this problem in a very neat way. What he did was to make up an *imaginary* father-figure – which is the idea of God. He endowed this father-figure with all the properties that he'd come to value so much in his earthly father. So he could look to him for love, provision, help, kindness and all these things. And therefore he had his cake and ate it at the same time. This, said Freud, is why the Christian idea of God bears so many similarities to aspects of what it is to be a human father. It's a very ingenious theory.'

Malcolm Jeeves then went on to describe how this line of argument could be turned on its head: 'The problem, of course, is that you can then say, "How was it that Freud became an atheist and developed that theory about why *other people* believe in God?" You could then look into his life-history. You discover that Freud had a very bad relationship with his earthly father. He had a Roman Catholic nanny whom he didn't like at all. And therefore, if you applied Freud's own theory, you would predict that he would do two things. He would want to get rid of any dependence on his earthly father, and he would want to get rid of religion. In other words, it's good clean fun thinking about these sorts of theories, but you have to take the whole issue much more seriously and ask what is the relevant evidence? Most of this, as far as I can see, Freud did not take into account.'

Carl Jung

What have other psychologists had to say about religion? Fraser Watts again: 'Well, of course, there have been psychologists who are a lot more sympathetic to religion than Freud. Carl Jung is the person who immediately comes to mind. But actually Jung is not *entirely* a friend of religion. What he calls "religion" is so maverick – at least from an orthodox position – that one has to take what he says with a pinch of salt.'

'You have in mind Jung's idea that there is a dark side, a sort of evil side, to God? Is that what you mean by "maverick"?' I enquired.

'Yes, that's right, the quaternity: the three Persons of the Trinity, and then (he thought) there had to be a *fourth* part, which he often saw as the Devil (though sometimes he puts the feminine in that fourth quadrant).'

Unlike Freud, who was mostly concerned with early childhood experiences and their knock-on effects for later life, Jung (who was a colleague of Freud's before they fell out) concentrated on the *latter* half of life. What primarily interested him was the process by which, later in life, we come to fulfil those potentialities we had earlier had to neglect through force of circumstances. He was interested in the development of the fully rounded, mature personality. He recognized that just as a round *physical* object has to have a centre, so the fully rounded psyche had to have a centre of some kind. And as far as he was concerned, that centre was intimately bound up with religion.

But just how important was religion to Jung's thought?

Fraser told me, 'He took a very broad view of what religion is. But he thought all of us had in our psyche symbols of what it would be like to be a whole person. These symbols, in our culture, are very often couched in religious terms. Christ, for example, is a symbol for us of a whole person. So he saw the path towards wholeness as being, in a sense, a religious one – a path towards making a reality for ourselves of these religious symbols which are in the psyche.'

'So you seem to be saying that a religious attitude, or a religious approach, is in Jung's thought central for what it is to become a whole person,' I said. 'But that's all happening in the mind. It is presumably independent of whether this religious experience relates to something which exists objectively – to a God. Did Jung actually believe in an objective God?'

'He was slightly shifting in his views about this, and it isn't really clear. I think what he's saying most of the time is that he has direct knowledge of the *symbols* of God in the psyche. But whether there is a metaphysical God is simply not a matter for him as a scientist. In his clinical work, he has knowledge of the *image* of God; I think that's what he meant.'

Anthony Storr had this to say about Jung. 'Jung was a great contrast to Freud, in that he was brought up in an extremely religious household. His father was a parson of the Swiss Reform Church. Many of his relatives and his wife's relatives were parsons.

'To his horror, when he was young he started having a series of dreams which clearly questioned the kind of orthodoxy in which he had been brought up. When he raised doubts about religion, his father (who was a rather conventional man) wouldn't argue with his bright son. And Jung got thoroughly disillusioned with conventional religion. But unlike Freud, it left him with a huge gap. Perhaps the whole of his psychology is trying to discover a way of filling that void in his life. Certainly religion, in the widest sense, was enormously important to him. He thought of many of his experiences as religious. He thought that if you contacted the unconscious, you could, as it were, find the voice of God within. I remember once when I saw him, he said about dreams that every night you have the chance of the Eucharist. He really thought that God spoke to him directly in dreams. He had a lot of ideas of that kind. To me, they all add up to a man who had never got over his loss of faith in early life. He was always trying to find in psychology – in *his* version of psychology – a religious faith of a non-doctrinaire kind, something of vital importance to him.'

I put it to him, 'So you have this contrast between Freud looking upon religion as a kind of neurosis (a sign of mental ill-health and immaturity) and Jung seeing it in a totally different way, as a sign of mental health.'

'Yes, Jung thought that nobody could be really healthy unless they had achieved a kind of religious attitude. This didn't mean believing in one particular set faith. It didn't mean being a Buddhist or a Catholic or whatever. But he thought that you could, in the course of psychiatric treatment, develop a religious sense, in that you could rely on a power within that was greater than your ego. Whether he thought that God was outside or inside is a moot question. I now think he did have a belief in a God outside. When he was interviewed by John Freeman on television years ago, he said, "I don't need to believe; I know." He felt that he was the carrier of direct religious experience.'

Leon Schlamm of the University of Kent is an expert on Jung. He explained the centrality of religious experience in Jung's thinking.

'He focused on the immediacy of religious experience and rejected the authority of religious belief and dogma. For Jung, what was authoritative was the immediate psychic (i.e. religious) experience.

'He was also very concerned about an issue which has been rather neglected in the Western religious tradition: the relationship between spirit and instinct. In the Christian tradition, they have been typically understood to be at loggerheads with one another. But Jung regarded the relationship between spirit and instinct as one of the primary issues in analytical psychology – indeed, the *collaboration* of spirit and instinct. He argued that it was possible to transform the energy of instinct so that it could actually work for spiritual purposes. He also felt that the Christian tradition had tended to put too much of a distance between matter and spirit. Matter has to be integrated into a spiritual world.

'Jung had plenty to say about God. For Jung, what is rather distinctive about God is that he has a shadow, just like man …'

'A dark side to his character?' I asked. 'He is not all good?'

'He is not all good. He is evil as well as good. This is rather shocking. For Christians, understandably so. But this was something that Jung began to play with in his late thirties, and it came out rather dramatically in his fifties, quite unequivocally. So God has a shadow, and he is therefore responsible for evil, just as we are.'

Religious experience

With 'religious experience' playing such an important role in the psychoanalytic study of the mind, I was prompted to enquire more deeply into what exactly we mean by the term.

I asked John Habgood how we experience God. He told me: 'Frequently people experience God through his absence, in a kind of restlessness and dissatisfaction; the kind of searching which is so characteristic of human life – the sense that there must be something more. St Augustine said this years ago: "Thou has made us for thyself, and our hearts are restless until they find rest in thee." For some people, these are the first warning signs that there is another reality they've got to come to terms with.

'Others may come to God through being blessed with a great deal of happiness and joy; they simply want to say "thank you" to something or somebody for what they are and what they experience.

'Others come to God by much more tortuous ways – frequently through suffering a tragic loss, a major change of lifestyle, or whatever. These events suddenly make people think, "What's the meaning of it all?" I frequently meet adults who have come to faith in later life. Almost always it's an event of that kind which has, as it were, broken through the crust of self-sufficiency. They began to see their lives as part of something much larger.

'Others may come simply by learning about Jesus Christ and saying "There's somebody who knew the meaning of life".'

How does a consultant psychiatrist regard religious experience? Montague Barker approaches it from a Christian perspective.

'I think it is where God meets us, and how we react to that in a psychological way. The way in which we react will depend upon our culture, our own background, our own personalities and the influences around us. This is part of the problem. When people have this experience, whether it's a funny feeling in their tummy or whether it's a vision, they feel that *that* in itself is a proof. I don't think it can ever be a proof of anything, except that you've had the experience. It's only if something comes out of that – in terms of changing your life, your whole motivation – that you can say yes, this is more than just a rumble in my tummy or some psychological projection.'

'So am I right in saying that, to some extent, you would go along with Freud in suggesting that a religious experience has quite a lot of yourself in it? But you would differ from him, dare I say this,' I added, patting The Couch upon which we were sitting at the time, 'by saying that this was nevertheless a reaction to an *objective* God which is other than you?'

'Yes. I really do agree with many of his views about religious people, religious experiences, and the way people react. But there is more than just the "nothing but" that he used. And it's that other which you then have to examine and ask where it comes from.'

I asked him where, amongst his reasons for believing in God, he placed religious experience, as opposed to other ways of learning about God. In view of the fact that he had spent most of his working life probing into the nature of mental experience, I was somewhat surprised by his answer: 'Maybe this reflects the kind of person I am and the kind of job that I do, but religious experience as such I would put way down. So many people come with experiences on which they have constructed all sorts of theories and attitudes that have got them into trouble. I sometimes have to unravel this feeling that's got them into an absolute mess. Feelings I don't deny, I don't reject. I just say, "We have to look at those feelings a bit more closely to see how valid they are for basing your life upon them." So I put religious feelings, as such, low down. But I would not devalue them in terms of the motivation they may give to somebody.'

'If religious experience is well down on your list of evidence for God, what then would you give priority to?'

'When it comes to a view of God, I think you look for a coherence in the belief structure. The structure of belief is more important than just the experience. But a structure that doesn't issue in some kind of experience and action thereafter is, of course, worthless.

'So there has to be a belief structure, and that comes from whatever revelation of God we meet. Many people have an experience of the beauty of nature. I believe that comes from a God who has created the very beauty that we respond to. Some people would say, "I don't like the idea of a God out there; beauty is something that comes from *within* us." But the fact that it is universal – that so many people respond to beauty – may indicate that we are created in the image of a God who creates beauty.'

Monty touched on the fact that for some people religious experience can be detrimental. One might, for example, find someone who has the problem of evil out of perspective, and becomes wretched and debilitated by an overwhelming sense of guilt. I posed this to Fraser Watts; did he think that the state of being religious was mentally good for you or bad for you – or was it a mixture of the two?

'I think it's a mixture. There's been a vast amount of empirical research on that. It depends on what sort of religion you have. You can make a rough distinction between people who are really committed to their faith, who seem to have slightly better mental health than the average population, and those people who have a merely nominal faith, who seem to have slightly worse mental health than average.'

I asked him how he would define or describe religious experience.

'I suppose it means different things to different people. But often what people mean is a general awareness of a spiritual dimension of life – of something beyond themselves that they usually interpret as benign and caring.'

'Do you think that there could ever be a systematic, scientific study of religious experience?'

'Oh, certainly. There has been quite a lot of research of that kind. You can identify, with questionnaires and in-depth interviews, what kinds of religious experience people have had. And there are certainly questions about it you can address scientifically: what sorts of people have what sort of experience; what effects it has on their lives. So there is certainly a scientific perspective on religious experience. But equally I think science doesn't have the *only* word to say about religious experience, or anything else. There are always other aspects to be put alongside the scientific one.'

'How important do you see religious experience as a route to understanding God?'

'Well, I think it *is* important, though there's been a division of view about that in the Christian tradition. There are really two things to be held in balance. On the one hand, it's important to emphasize the tradition and practices we inherit in the Church. But I also think it's important to emphasize the personal and experiential aspects of the religious faith. I want to see a to-and-fro between those.'

The organization best known for carrying out systematic studies of religious experience in the UK is the Religious Experience Research Centre at Westminster College, Oxford.

Its current Director, Laurence Brown, explained its background: 'This centre was established twenty-five years ago by Sir Alister Hardy, who was a professor of zoology at the university. He believed very strongly that religious experiences were in some sense given to us by God, and that they were a way in which God disclosed himself to people. In order to support that theory, he placed advertisements in newspapers asking for people who had had an experience that they thought took them out of themselves. Whether they called it "God" or not was unimportant. These experiences have been collected into an archive that the centre holds. There are now about 6000 accounts, with the letters that accompanied them.

'If you look at these accounts, it's very clear that people feel that a religious experience does just what Alister Hardy thought it did – it gave them a perspective (some other view about the world) different from what everyday life was about.

'People have a lot of experiences (perhaps when they are on the tops of mountains) where they are impressed by the grandeur of nature. Also, we know that people have very powerful religious experiences within church contexts, or in their relationships with other people. There is some sense of the transcendent. They are transcendental experiences because they give the idea of there being something that is other than, and apart from, us as human beings.'

'Has the centre done any systematic surveys of what proportion of people have these experiences?'

'Yes, about half the population (women perhaps a bit more readily than men).'

'Does a centre like this simply collect data, or do you try to form theories about what actually is going on? Do you think these experiences refer out beyond themselves to a real God?'

'The last of those questions is probably impossible to answer. How would one know whether the experiences people have refer to a real God in some objective way? The most one can say is that if a person has an awareness of God, and that changes their lives, then "God" is a real influence on that person. A lot of people are able to draw attention to a very specific event – a single occasion – which changed the direction of their lives. But I guess there must be a lot of other people, like me, who have always lived in the shadow of some transcendental order.'

I next asked theologian and philosopher Keith Ward the same question I had earlier put to John Habgood: what he would say were the ways in which God is able to make himself known to us?

'I think God is known in very ordinary ways, not necessarily anything extraordinary at all. It's more like a general spiritual presence, a feeling that everything around you, your whole environment, is shot through with the presence of a conscious being of some sort. I don't think the world needs to look any different; it's just a special way of discerning it – as a spiritual world, not just a material one.'

'Having sensed the presence of God, are you ever going to be in a position where you can say, "I have proved that God exists"?'

'Proof is very difficult. I have taught philosophy for thirty-five years and I've never proved anything yet! In fact, I seem to *dis*-prove more things as I go along, or at least find good reason to doubt them. So I wouldn't think in terms of proof.

'Instead, you're thinking about what you commit yourself to. It's like committing yourself to some ordinary human person. You wouldn't talk about "proof". You'd just say, "This has become the basis of how I live." I think it's like that with God. You as it were *bet* your life on there being this reality of beauty and truth which calls you on towards it. And that's the way you live. And after a while, you can't see any alternative really. But it's not a strict proof.'

So one senses the presence of God. But can the religious experience be more specific than that? Can God actually communicate: send messages to us, tell us what to do, answer prayers?

Astronomer Jocelyn Bell-Burnell, speaking from her experience as a key figure in the Quaker movement, thinks this is so: 'I certainly don't think God steers the clouds away from the garden fête! I don't believe that God intervenes in that kind of way. I do believe that faithful people (God-like people, God-attuned people) are forces for good in the world. God works through people. I don't think it's a crudely mechanical thing. I don't think God says, "Russ, go and visit the lady next door this afternoon, please." I doubt if it's that directed. But I think those of us who endeavour to be faithful, good people – not that the only good people are godly people – are the ways in which God operates in this world, and I suspect they are probably the only way God can operate in this world.'

'But how do you see him influencing people and getting across his ideas?'

'I don't see him as "him",' she gently admonished me. 'A lot of the qualities of God are more traditionally labelled feminine, so I try not to be too specific about the sex of my God.'

'Fair enough.'

She continued: 'In the Church that I belong to, the Quaker denomination (the Society of Friends), we gather on a Sunday morning in silence to wait and listen, and we are enjoined to be alert to the promptings of love and truth in our hearts which are the promptings of the Lord. So we gather in silence, and out of that silence anybody can speak. Few do, it has to be said (it's not bedlam!). And we listen very hard to what people have to say, and we listen very hard to the silence within us. One is looking for a nudging – the same kind of feeling as I would attribute to intuition, that sort of gentle, small prompt. And sometimes one can be quite consistently nudged in a particular direction. You become sensitized to an issue, and you find your mind turning to it more and more often. And you notice things in the press about it or what have you. And gradually you will develop a concern for whatever it might be.'

'Can I be clear?' I persisted. 'In your meetings you have these nudgings, these thoughts being stirred up in your mind. Is this purely subjective, or do you believe that these nudgings are coming from an existence which is other than yourself – in other words, from God. If you did not exist, would your God exist?'

'I'm sure some of these nudgings are very subjective. "Bees in one's bonnet" is another way to describe them. But one of the arts of working in a Quaker group is to distinguish between bees in bonnets and what we believe to be the leadings of God. So we work on the assumption that there *is* a God, and that God can give us leadings through these nudgings. Whether we say God is "beyond us" or "within us" is difficult. I think the answer is both. I certainly believe that God does not die when I die.'

That, then, is how it can feel. But how does prayer actually work? How does God get his (or her!) thoughts into your mind? Keith Ward had something to say on this which I found very helpful: 'God is omnipresent. God's not excluded from anywhere. Wherever you are, God is. God's not just outside you; God's actually *inside* as well. You're part of God, and God is part of you. So it's not that there's another being there, with different thoughts, somehow trying to communicate with your thoughts. It's that God is *already* in your thoughts. Prayer is very largely a matter of coming to realize the presence of God within. Nobody is in a totally private world, because you're always part of the life of God. And that opening out of your solitude into a wider communion of being is really what prayer is.'

*

95

So how are we to sum up our present-day understanding of the relationship between psychology and religion, particularly that aspect of religion known as religious experience? Where did Anthony Storr see psychology as a field of study progressing?

'Over the last twenty or thirty years it's become enormously more open-minded than it used to be. We had quite a long period of being locked into a narrow behaviourism when it was almost impossible to investigate what most people call "the mind". We have now moved beyond that. The range of things that psychologists are happy to investigate is enormously broader. There is a bit of residual narrow-mindedness in some quarters, and interestingly that is often associated with attacks on religion. It is often said that human beings are nothing but this or that: nothing but our nervous systems, nothing but a survival mechanism for our genes, and so on. I don't think that kind of narrow-mindedness has any scientific basis, though unfortunately, as I say, it's often used as an attack on religion.'

'So would you say it's becoming more respectable for a psychologist to have religious views?' I asked.

'Yes, I think so, and I know a lot of psychologists who do. There is as great a diversity of people in psychology as there is in every other science.'

'How would you regard psychology as impacting upon religion today? What do you think psychology does to help us to understand religion?'

'I think there is a great move on to reconcile the two. Because of Jung's work primarily, people have begun to realize that you can trust something within yourself, or trust a higher power of some kind, without having an orthodox faith. That is a very valuable idea. I would have thought more and more people would take to that point of view. So I don't think that modern psychoanalytic psychology is in any way necessarily opposed to a religious point of view. Certainly, when I was in practice I didn't have any difficulty in treating people with religious beliefs which I didn't share.

'But of course, when we talk about "psychology", we're only

talking about a very limited kind: the psychoanalytic psychology of Jung and Freud, and so on. You've got to remember that the bulk of psychology in this country is now done by experimental psychologists – those who treat human beings as experimental animals and keep their distance from them. They are not primarily interested in the kind of things that Freud and Jung were.'

'What insights do you think your lifetime of studying psychiatry and psychology has given you into what it is to be a human being?'

'What an impossible question! What is it to be a human being? I don't know. I think one result of practising as a psychotherapist and teaching doctors has been to enlarge my capacity for getting in touch with a variety of different kinds of human being. I'm impressed with the *variety* of human nature. We mostly live in rather small circles of society. But if you're a practising doctor, you have the opportunity of meeting people from all walks of life. They may hold quite different opinions from yours, hold religious opinions that are quite alien, come from different countries and different social backgrounds. I think that's enormously valuable. It does give you an insight into the richness of human nature and its diversity.'

Neurosciences

The trouble with investigating the mind is that it's so intangible; it's difficult to come to grips with it, and impossible to be quantitative about it. That's why it seems so attractive to deal with something physical like the brain. There is clearly a close link of some kind between mind and brain. So is it possible that a surer route into the mind is through the neuroscientist's study of the brain? I asked Malcolm Jeeves where he saw the cutting edge of psychology these days.

'I think the most exciting parts are where it is impinging upon other disciplines – the bits that link up with neuro-physiology, genetics, biochemistry and that sort of thing. To be more specific, the areas concerned with trying to relate what's happening in the mind with what's happening in the brain – the so-called mind-brain link.'

This seemed to be another area that impacted directly upon the religious outlook. Does such work hold open the possibility of dissecting what it is to be a human personality? Can the sum total of what it is to be you or me be expressed in terms of the physical structure of the brain, its chemical flows and electrical currents? I asked Professor Jeeves what methods are used to determine which parts of the brain are associated with which mental experiences.

'There is a variety of methods,' he said. 'The classic one is to look at people who, for whatever reason, have suffered damage to their brains – through accidents, or through being born with unusual brain structures. You can then find out the way they think, the way they see the world, the way they remember things, and how these differ from those of people who have normal, intact brains. That gives you one lead into it.

'Of course, you don't really know precisely how much the brain has been damaged. And even if the brain ultimately comes to autopsy, you can't always be sure what other parts have also been damaged. For this reason, one of the most exciting alternative ways of studying it is by doing very carefully controlled surgery on non-human primates. Here you can very specifically limit the size of the brain lesion you make. Then you can have a much more accurate account of precisely what you damaged. And because animals are so similar to us in so many ways (the way they are conditioned, or they learn, or solve problems, or they remember), this means that we can, in a very much more scientific way, check up on the findings we've already got from studying these brain-damaged people I was talking about earlier.'

He continued, 'There is another dimension to this which, in many ways, is the most exciting for some people. This is the attempt to link the changes in the chemistry of the brain with what is happening in mental life. We all know that if you have slightly more alcohol than you should, then the state of your mental processes changes. But we also know that various other drugs do this. The helpful and exciting thing is that, with the advances in pharmacology, one begins to see how quite sophisticated changes in particular chemicals in the brain are then reflected in changes in the way a person sees the world, or remembers, or thinks, or whatever. So that is a very important approach these days: to link the neurochemistry of the brain with the psychology of the person's behaviour.'

I asked him how he saw the relationship between brain and mind. Did he regard them as two different entities connected in some way, or is the mind simply the brain seen from the inside, or what?

'The obvious and natural way is to think of the mind and the brain as two different things, two different substances. This is understandable. However, when you think about it a bit, and look at the evidence for the tightening of the link between mind and brain, then this so-called "dualist substance" view becomes to my mind more and more difficult to hold. The link between mind and brain is so close, so tight, that I think to see them as two separate substances is not very helpful. When I say the "tightness of the link", I mean this: As I'm speaking now, you could describe what's happening in terms of my attempts to formulate answers to your questions, to put together sentences, and to give you some logical thoughts. For the purposes of what we're doing now, that is the most important way to describe what's happening. On the other hand, it would be perfectly possible, if you chose to do so, to wire me up to an EEG machine, put my head in a positron emission tomography machine, and look at which parts of my brain are active when I'm answering particular questions. What I would argue is that to give any complete account of what is happening, you need to include both of these.

'Now if you were a thoroughgoing reductionist, you would say that what is happening is nothing but the activity of my brain. But the problem with that is that the assertion of such a thoroughgoing reductionist would *itself* be nothing but blips in the neuronal network of his brain. In other words, he wouldn't be saying anything that was open for rational debate – it makes nonsense of things at that level.

'So my view is that there are two essential aspects, the mental aspect and the physical aspect, and that both are necessary if you're going to do full justice to the totality of what is happening. The usual analogy is with writing with chalk on a blackboard. If you're a good chemist, you should be able to give a complete account of everything that was on the blackboard in terms of calcium carbonate and so on. But you would then have said nothing about the fact that the chalk had written on the board something like: $2 + 2 = 4$. But that's the most important thing it's trying to communicate, not the make-up of chalk.'

'How do you interpret what the Bible says about body, soul and spirit? Does this need to be brought up to date?'

He thought for a moment, then said, 'We are living beings, as are the animals. But we also have a capacity for a personal relationship with God. This tends to be described as the *spiritual* aspect of our being. I don't regard it as another thing that gets tagged on somewhere. I regard the spiritual aspect of a person as the operational definition of who he or she *really* is: what they give their highest priorities to; what their values are; how they behave towards other people, towards God, towards the creation. If you put all of these things together, they will indicate something about the spiritual nature of a person. It is a dimension of their being.

'The diversity of words used in scripture to describe human beings – whether it's soul, body, spirit or whatever – are all attempts to do justice to the many different aspects of the full life and functioning of a person. They're not separate things which are then glued together in some way or other.'

This journey of mine being about science and religion, I felt it was high time to pay a visit to a scientific laboratory. You will recall that during my earlier conversation with biologist Steven Rose, we were in his study and separated by a brain floating in a glass tank. This time I accompanied him to his lab. There he does experiments on the brains of animals. He uses newly hatched chicks. I asked him what was the motivation behind the experiments he did.

'What I'm interested in specifically is the way the brain stores new information. Memories are stored in the form of changed connections between the millions (even the billions) of different nerve-cells within the brain. What we are doing is looking at the ways in which these connections are made, and broken, when the animals we work with learn very simple tasks. We work with very young chicks because they learn a great deal, very fast, about their environment. But I am pretty sure that the sort of changes that take place in the chick brain are very similar to the things that go on in your brain, or my brain, when we are learning and remembering things.'

'Is your hope that one day this will lead to a complete explanation of personhood?'

'Well of course, it depends on what you mean by a "complete" explanation. Also, I should say that I don't really think of the brain as simply a collection of molecules or a collection of cells. A brain is essentially dynamic. It has structure. It's the structure, the connections, the pattern of relationships between the individual cells and the electrical flows, which are so important. You have to consider it as having a history of its own. Our brains are what they are *now* because of the changes – all the molecular and historical events – that have taken place during our development.

'Of course, I am a materialist. I do think that the mind and the brain are, if you like, identical. What we have are two different *languages* to describe the same phenomena. When I work on the molecular changes which go on during memory formation, I am also in a sense tapping into changes that are going on in the minds of my chicks. I believe that it's possible to find the ways of translating between the biochemical and physiological language that I use for the brain and the language that we use for the mind. I work on memory precisely because I think it's a way of uncovering what I would call the "translation rules" for the languages we use to talk about mind and brain.

'Now, will that tell us what it is completely to be a human being? No, I don't think so. I might be able to translate the phrase "I am angry" or "I am in love" into statements about the electrical properties of particular cells or the molecular changes that are going on within them. But there is something about the quality of being angry such that we shall still want to use the language of "anger" and not the language of "cells".

'So I think that those of my colleagues who are neuro-scientists and who are what I would describe as "hard-line reductionists" (that is, they want to get rid of the language of mind altogether) are fundamentally misguided. We need these languages, but we need to understand that there are, if you like, lawful relationships between them. We are in the business of being translators – not interpreters, not reducers.'

Biologist Will Provine would not exactly classify himself as a 'hard-line reductionist', but he is nevertheless very optimistic about the final outcome of these physical experiments on brains. He told me how he believed we shall eventually understand how material minds produce consciousness and awareness.

'I look forward to a time in several hundred years, or perhaps even sooner than that, when scientists would be able to understand how to take a combination of a biological system and a computer small enough to fit into a child's toy so that the child could enjoy a toy which has consciousness and awareness.'

'Are you saying that a human being – the sum total of all that we are in body, mind or spirit – will one day, in your view, be completely described in purely physical terms as atoms and their movements and their forces?'

'Not really. I can't go with that kind of reductionism. The reason is because reductionism in any biological organism can only be down to the level of the ordered molecules that were inherited in most organisms – in other words, to its DNA. That DNA has a long history of three and a half billion years of biological evolution. So for that reason, I do not agree that we can ever reduce human beings to just physics and chemistry.'

'But you seem to be saying that one day a study of the brain is going to give a complete understanding of consciousness. And that study of the brain will be in terms of matter, electricity, chemicals, and things of that kind. It does seem to me that you have reduced everything we think of as being a human being to simply the movement of matter.'

'Again, I don't agree. I think that we will reduce consciousness and awareness to the activities of *living cells*. I suspect that the consciousness we will ourselves be able to create will involve the activities of *living cells* as part of the machinery. I don't agree that we're likely to reduce human thinking purely to physics and chemistry.'

'So are you saying that there's something intrinsic to the living cell that cannot be explained in terms of the movements of atoms?'

'That's exactly what I'm saying. What can't be explained about a living cell in terms of pure physics and chemistry is the part which came from its hereditary history. In order to get a good physical and chemical explanation of the activities of any one cell, you would need the entire biological history of the DNA that went into that cell. So I just don't agree that that kind of reductionism to pure chemistry and physics is possible.'

I was not convinced. I still had this lingering feeling that, although it would admittedly be a mammoth task to assemble all that historical data, in principle it could be done. And therefore, it seems, the conclusion one is led to – if you go along with what Will Provine was saying – is that, in principle, it would be possible to reduce everything to just physical entities and their movements.

Someone with an avowed triumphalist attitude to science is Peter Atkins. He was in no doubt about where neuroscience was heading.

'Science will, in the future, be able to resolve what is meant by consciousness. It will resolve what it means to believe in things, to think that something is beautiful, and so on. This, of course, includes the nature of religious belief and the nature of aesthetic appreciation, all that sort of thing – the things that we lump together as "human spirit". I also think that it's almost inevitable that science will be able to build machines that can simulate consciousness so completely that we won't be able to distinguish it from the consciousness of you or me.

'But we won't, I think, be able to predict the outcome of a conscious being. The ability to make predictions of extremely complex phenomena will not be within the domain of science. After all, it's not possible at the moment even to make predictions about very simple phenomena, like one pendulum hanging from another and the way that it swings – an absolutely trivial problem. But that doesn't mean we don't *understand* this problem of one pendulum hanging from another. So I think that in due course we will be able to simulate consciousness, build machines that are conscious, and in a sense understand everything there is to know about consciousness. But we will not be able to predict that if you tickle this machine with a feather, it will produce a sonnet.'

Theologian Keith Ward could not disagree more. In response to my question about whether he thought that consciousness would be explained one day, he declared, 'Absolutely not!'

He added, 'That might sound arrogant from a non-scientist, but the really important things about human life are how things feel to you, how you see things, what meaning and significance you find in them. Talk about physical particles doesn't even mention things like that. The best you could do would be to establish some correlations between the physical events and what somebody tells you when you ask them "What are you thinking when you see these events?" And even if you establish those correlations, there will always be people who have new thoughts, who do new things, and who have new and different sorts of experiences. So a physical study of the brain alone is never going to catch up.'

Roger Trigg, philosopher from Warwick University, adds his own reason for being sceptical of the more optimistic claims.

'I doubt you could ever get a complete scientific understanding of human consciousness or of the human mind. After all, who's doing the understanding? It is a human mind that's doing the understanding. So I feel that people who are trying to explain the mind are in the end sawing off the very branch they're sitting on! They're explaining – indeed, explaining *away* – the very thing that enables them to participate in science in the first place. It's a bit like somebody drawing a picture of a room in which the person is drawing a picture. You never catch up with yourself. Consciousness, therefore, can't wholly explain consciousness, just as science will never completely understand science.'

John Habgood had a similar reason for being cautious about how far brain research can reveal the real you and me.

'We don't completely understand anything at all,' he said, 'let alone the mind. The whole process of understanding is trying to peer deeper and deeper into what is ultimately incomprehensible – and that is true of the whole Universe. We can draw up marvellous plans of bits of it. We can say that we do understand the relationship between X, Y and Z. But when we come to self-understanding (which is what understanding the mind completely would entail), then I think we begin to reach a logical paradox. It is the characteristic of our human self-consciousness that there is always that something behind the knower which is ultimately mysterious.'

Arthur Peacocke attacked the very idea of trying to explain complex systems in terms of their component parts.

'I think this is a typical example of how a more complex system generates, though its internal relationships, complex behaviour which *in principle* is not reducible to the concepts that describe the bits that make it up. A radio set can do all sorts of things that the chips that constitute it can't do if they are in a heap on the floor. The human brain is the most complex piece of matter in the known universe. We shouldn't be surprised that it has activities which are not reducible to descriptions of the movements of the atoms, and the firing of the neurones, that make it up.'

What this seems to add up to is that, while science can investigate all the physical aspects of our brains, there is still something about the mind – and therefore about who you and I really are – that science can't get at. It's not a gap in scientific knowledge, it's not a missing piece of a jigsaw puzzle. It goes deeper than that. If we are to do justice to our subject, we seem to need more than one language.

*

Theologian and former physicist Ian Barbour takes up this theme of complexity, and introduces the idea of different levels of explanation.

'In the tradition, we've perhaps been guilty of somewhat dualistic thinking, setting mind (or soul) over against the body. This goes back to Greek thought. The Bible itself is actually closer to science at this point in seeing the human being as an integral unit.

'This is where thinking in terms of levels helps. You can think of higher levels in us as being distinctive, as involving relationships that you don't get at lower levels. Sure, we're made of atoms, and we're made of cells, and we're made of organs. But these are put together in total interacting *wholes,* in which types of phenomena can occur that you can't describe in the lower language. I mean, you can't describe a thought, or an ideal, or human love, or purpose, in the language of physics and chemistry. That doesn't mean that we're not atoms. But we're a lot more than atoms. Things go on in us for which you need distinctive concepts to be able to talk about them.

'We're selves in community. We're not just isolated individuals or isolated minds; we're part of a social order. We're constituted by our *relationships.*'

So where exactly should we locate 'personhood'? Is it rooted in the mind, or is it to be found in the body, or is it a combination of the two? I had thought that Montague Barker, being a psychiatrist, would have tended to identify it with the mind, but not so.

'I have to be honest and say that I can't think just of pure mind or pure body. I think of people as being embodied minds, or souled bodies. When I'm meeting people, I'm meeting a physical person. That body interacts with their mind. The way they think about themselves is altered by their body. The way they behave is very much physical, as well as something that they're thinking in their mind. Sometimes the physical bits run away with their mind, and the mind has occasionally to say, "Now just a minute, that's gone too far," or whatever. This interaction is the way I would see it; I don't like to split off body and mind.

'You ask me to make one more important than the other. Obviously I'm dealing with the mind. But the way in which people see their body, the kind of genetics they have, the kind of life they've lived, the way they've handled their body, these also affect the mind very deeply. I can't just dissect that off in the way you suggest by your question.'

Designer drugs

So we are to see the body and the mind as an integral unit. Though we may need different levels of explanation to get a full picture of the real you and the real me, there are certainly very close links between those levels. And nowhere is the connection between brain and mind made more apparent than when we see the effects of taking certain drugs. These days, there is talk of 'designer drugs'. John Cornwell is a philosopher by training and Director of the Science and Human Dimension Project at Jesus College, Cambridge. He has made a study of these new drugs and their possible future implications.

'Let's be clear first of all about the difference between "designer drugs" and other kind of drugs,' he began. 'Until quite recently, most drugs which affected the mind were what pharmacologists called "serendipitous". In other words, they were discovered by chance. The difference with designer drugs or, as they are sometimes called, "rationalized" drugs is that since the great revolution in molecular biology in the seventies and early eighties, it's now possible to design compounds which affect receptors on nerve cells in a very specific kind of way. They know what they're doing, and the effects are more clean – in the sense that there are fewer side-effects.'

'What kind of changes in human behaviour, attitudes and feelings do they produce?' I asked.

'The effects tend to be within the areas of disease and mental illness. So it would be a mistake to think that rationalized drugs are in any way enhancing healthy human nature. We have to start with the proposition that they are having an effect on some known *defect*, for example an excess of a certain kind of brain chemical. So it gets the brain back to proper working order. That's the concept behind it.'

'I know a person who is schizophrenic,' I said, 'and another who is paranoid. These people find it almost impossible to operate on a normal daily basis without the help of drugs. Once they are on drugs they behave in a more or less normal way. What bothers me about these people is: who is the *real* person? Are we to regard the "real" person as the one on drugs behaving normally – or is it the person who cannot cope but who is not taking any artificial drugs?'

'That's a very interesting and also a very difficult question. Probably the best answer is to say that you often see in people who are mentally ill the essential humanity or personality trying to exert itself through various sorts of excesses and deficits. I am sure that there are drugs which help people to overcome certain kinds of basic physiological difficulties, in order that their essential humanity can exert itself. But I would be very unhappy about the idea that there was a simple *equivalence* between the chemistry and the personality. One of the great heresies of the modern age is the notion that there is a direct equivalence between levels of brain chemicals and human identity – that we're just a sort of chemical factory in the brain. Once we go down that path, it means that we fail to see on the one hand the importance of a person's history as an individual (the history of one's relationships), and on the other, one's essential humanity (in the sense of the soul) which is, in my view, the whole of a person, including the whole of their bodily existence.'

'So you don't see this research into the effects of drugs as denigrating what it is to be a human being?'

'It seems to me that the crisis lies with the perception, in mass culture, that we are simply a balance of chemicals. It's a very easy and simple view of human nature. It's the one that seems to be gaining ground in the culture itself, and of course there are social and political arguments for adopting that particular viewpoint: the idea that if human beings are a concatenation of chemicals, they would perhaps be more controllable.

'But I am not at all sure that many scientists working in the field would claim that human identity is simply a matter of brain chemistry – though there are the more reductionist scientists.'

'Do you foresee the possibility of a future where these designer drugs could change our behaviour so radically that we might all become very kindly and tolerant? Would it be possible to manufacture a society which was all on pills designed to make people into good citizens?'

'I think that's an almost irresistible idea. Let me give you an example: in the United States at present, I am told that something like 32 per cent of all children have been diagnosed as suffering from a condition known as "attention deficit disorder". This means, in the minds of many parents and physicians, the sort of child that comes home, throws its coat on the floor and goes and watches television – instead of doing homework. And there is a pill for this, which has been quite massively taken by millions of children across the United States.'

'When they're on this pill, they will quietly settle down and do their homework?' I exclaimed, in some astonishment.

'They take their pill and become much more amenable to the sort of tasks that adults would want them to perform, yes. Here you have an example of a behavioural norm which has been dictated by a pharmaceutical company. Here are people telling us what constitutes a disorder in children. And I think that is extremely dangerous.'

'Suppose you had a pill which made everybody very well-behaved,' I suggested. 'Would that raise religious and ethical questions?'

'I think any notion that we can, as it were, *buy* moral behaviour, or buy religious experience, through consuming a drug is clearly extremely dangerous. In the first place, it impoverishes the human image. It seems to have very little to do with our status as a moral agency and our having free will. I can only imagine that if these drugs are used for these sorts of purposes, as opposed to genuine therapy (for illness and disease), we shall have arrived at Huxley's *Brave New World*.'

'Can I ask what you see as distinctive about human beings –

what it is that raises them above the level of being merely a chemical factory?'

'It needs to be said in any discussion of this kind that we have quite recently come out of a very long period of seeing human beings in a dualistic way, split between an immaterial soul and a physical body. Indeed, one of the great excitements in neuroscience, and in the studies of the chemistry of the brain, is the sense that we are at last putting the soul back where it belongs: in nature itself. We are finding that our souls are fully embodied, and that it's not a kind of spooky stuff floating around inside the physical body.

'Now I think we have to be extremely careful. What is happening at the moment is that there are people who are searching, if you like, for the soul through science; they think that the answer to everything lies in looking at the smallest parts – finding how the whole works by looking at its parts. This is often called "reductionism". Then there are those who are unhappy with that; they feel that we still have to see things in the round; we have constantly to return to seeing the *wholeness* of things. And that's often called "holism".

'My view is that, while I accept that the old, dualistic way is incorrect and leads us into many errors, it's not at all easy to come up with a *new* account of the human soul. In the first place, we have to hold on to the importance of seeing a human being as *history*. We are the story of the whole of our lives. Of essential importance too is the idea of the *imagination*. Philosophers, poets and thinkers down the centuries have tried to grasp what the imagination *is*, and how it works. It does appear that we are getting closer to this by the study of the brain and the mind, and even brain chemistry. It is through the imagination that we make contact with the idea of the creator, the idea of God. Imagination escapes the limitations of our finite nature. Understanding the imagination in a holistic way, as opposed to a reductionist way, is in the future probably going to be a vital key to reformulating what we mean by the human soul.'

'Traditionally, when we think about ourselves as not living our life the way we *ought* to be living it, we turn to religion (or perhaps in modern times, to psychiatrists or counsellors),' I said. 'Do you ever see a time when these drugs could possibly take over from religion, psychiatry and counselling?'

'I think it's going that way already. It's called by some "biological psychiatry". Instead of going along to a therapist or psychologist or psychiatrist, who would talk through the difficulties, you simply go along and get a pill as a quick fix. That way of treating problems and mental illness is irresistible because it seems to be effective, and of course is much cheaper. As is well known, effective therapy can take anything up to six or seven years, whereas a dose of a certain kind of medicine (one of the more famous ones is Prozac) can maybe take just two weeks to kick in. But as time passes, these rationalized drugs (which have only been on the market for five or six years) are beginning to show their problems. I think that in the fullness of time people will realize it's not so easy to get a quick fix that way.'

Artificial intelligence

We have been talking about one way of designing people to specification: the modification of their behaviour by drugs. Another possible way is to start from scratch. One takes pieces of electronic circuitry and builds exceedingly complex, computerized robots – artificial intelligence, or AI. How far can one go down this path? How close could we come to building a machine like a human being? And does the mere prospect of such a development have any bearing on how we should view ourselves?

To find out, I went to the Reform Club in London's Pall Mall to talk with Margaret Boden. She heads the Philosophy Department at Sussex University, where a colleague of hers, Paul Hodgson, a professional jazz saxophonist, has recently fathered a musical genius. It is a computer program that improvises jazz in response to the sounds that it hears. I once heard a tape of the computer playing with live musicians. I had to be told that the AI member of the group was the one on piano! I know nothing of jazz, but it sounded pretty good to me.

I began by asking Margaret how she got into AI.

'Ever since I was at school I've been absolutely fascinated by the human mind: how it works, its relation to the brain, and its relation to our evolutionary background. That's always been my prime interest. I'm interested in artificial intelligence because it gives us some concepts and ideas (and some methods for testing those ideas) that can help us to think about the human mind.'

'What do we actually mean by "artificial intelligence"? Are we envisaging a time when a computer becomes fully conscious – has feelings and thoughts of its own, and becomes aware of things?' I asked.

'Oh, you don't have to commit yourself to that if you're interested in artificial intelligence. Without getting into those very heavy areas, somebody might simply want to use artificial intelligence – that's to say, complex computer systems – to model the sorts of things that human and animal minds do.

'Of course, some people *do* want to explain those things. But having a theory of them, even if it's a theory that you model on a computer, isn't the same as actually putting those things into a computer program. An analogy people often use is that you can model a hurricane in a computer system, but you don't of course think that there are actual winds and wetness inside the computer!

'In the case of consciousness, it's more controversial. The reason is that, from the philosophical point of view, we don't actually know yet what we mean by "consciousness". Because we haven't got the concept clear, there is disagreement about whether there could ever be a computer (an artificial intelligence) that was fully conscious. Some people would say, "Yes, in principle it could", other people would say "No, in principle it never could be". This isn't a purely scientific question. It's largely a question of the philosophical interpretation of consciousness.'

'When you say we don't understand the concept of consciousness, obviously in one sense we do: it's to be aware of things and to feel things. I can fully appreciate that it's difficult – perhaps impossible – ever to understand why a particular configuration of atoms (our brain or a powerful computer) has feelings and thoughts, whereas another configuration does not. But surely, as human beings, we know what consciousness is?'

'I don't think we do know clearly. You said, for example, that we know what it is to be aware of something. Well, what *is* it to be aware of something? Is it, for instance, the ability to react in some vaguely appropriate way? If you're talking about a creature with language, is it the ability to report that it happened, to comment on it, and to describe it? That, I would say, is a very important part of what we mean by "awareness". And all of that could be put into a computer system.'

'Certainly a machine can be programmed to react physically to things, without itself knowing it's doing it. But surely there is a difference between a machine that simply behaves in that kind of way and ourselves, who not only react to external stimuli but can sense, can know, that we're reacting?'

'If you're saying you have a gut feeling that there is something else going on in you, over and above these judgements and actions, I have to confess I share that feeling. Some people argue that if you try to say there is something over and above every aspect of your behaviour – which could, in principle, be put in computational terms – then you're deluding yourself. Now I'm not saying I agree with that. What I'm saying is that what you and I think we're saying when we express that gut feeling is not something that anybody has so far made clear.'

Besides consciousness – whatever that may be – another aspect of what it is to be human is the fact that we have free will. Admittedly, it's difficult to understand where that sense of free will comes from, given that our bodies are physical objects made up of atoms subject to the laws of nature. Nevertheless, we all live our lives on the necessary assumption that we do have free will – unless, perhaps, you're a Will Provine. I asked Margaret Boden whether she envisaged a time when an artificial intelligence might have free will – when it would be able to make decisions of its own that have not been programmed into it by the maker of that computer.

'I think talking about free will is a way of discussing the indisputable fact that we have ways of controlling our actions which young babies don't have, and which, so far as we can tell, no other animals have. And they're very important. They're important because they're bound up with notions of value and responsibility and morals, and so on. Human minds have a very real cognitive architecture (if you want to use that term) which, so far as we know, no other minds have.

'But in principle at least, one could put that sort of thing into a sufficiently complex computer program. If you say, "That's all very well, but it was just programmed into the computer by some human being", then I would reply that in the sort of complex systems I'm thinking about, that isn't necessarily true – in two interesting ways.

'One is that human beings are all different and very complex. This is partly because we have had different experiences and we've learned different things. A computer system of the sort that I'm talking about – a learning system – would also have had its own individual history. So on a particular occasion, you wouldn't be able to point to a rule in the program and say, "Joe Bloggs, who wrote this program twenty years ago, actually wrote that in." There's absolutely no reason to think that he did. It may have been generated or learned from experience.

'And the second point is even more interesting. There are already computer systems which do what they do not because some human being has sat down and written a program telling them, "If this happens, do this, and if that happens, do that"; they have evolved ways of doing things. These have been evolved over many, many hundreds of generations of slightly varying (randomly varying) versions of the program. It doesn't have any rules or instructions in it saying, "This is the task you've got to perform, and this is how you perform it". Once again, you couldn't say at every point, "It's only doing this because the programmer built it in", because the programmer did *not* build it in.'

I moved on to another distinguishing characteristic of humans: creativity. Could we think of computers one day becoming genuinely creative, writing new music for example. What about the music produced by Hodgson's jazz-playing program?

'Paul would say it's not "creative jazz", in the sense of a musician who changes the style of jazz and who goes down in the history of jazz as having done that,' she replied. 'But if he were looking for someone to play with, and he heard a human being playing in that way, he would be perfectly happy to ask them to join him. In other words, it's at the level of a competent professional jazz musician.'

'What would you say was the source of creativity? How do we come up with new thoughts and new ideas?'

'Broadly, there are two different sorts of creativity. One of them is coming up with new combinations of familiar ideas. I mean what happens when people come up with new analogies, or when poets come up with new imagery. I don't think that novel combinations of familiar ideas are particularly mysterious.'

'A computer could be quite easily programmed to bring up these combinations?'

'Yes, certainly. Obviously not programmed in the sense of telling it to come up with such and such an image, but providing it with a network of units, and having the thing automatically recognize certain sorts of similarities which perhaps *we* haven't recognized.

'But there's another sort of creativity which I think is more interesting. The shock of novelty it can give you is much greater than the "associative" sort of creativity can. Perhaps the best way to describe it is as *styles* of thinking. It may, for example, be a way of thinking about chemistry. In the middle of the nineteenth century there was a chemist called Kekulé who, like a lot of other people, was puzzling about the structure of the benzene molecule. He thought, as everybody else did at the time, that it had to be some sort of string of carbon atoms with other atoms hanging off it. To cut a long story short, he had the idea of changing the string into a *ring*. Could there be molecules that had rings of carbon atoms? And it worked. He not only solved the problem of the structure of benzene, he in effect brought into existence a whole new chemistry, a new sort of molecule. Now they were able to ask questions like: What happens if there are different things at the corners of this hexagonal ring? If you have two rings stuck together on one side, how would you do that? and so on.'

'I can see how a computer might elaborate on this new chemistry. But do you think a computer could have actually come up with the original idea of the ring rather than the string?'

'I don't see why not. You could have a computer program which was built to worry about chemical molecules. It could have somewhere in it the explicit statement that organic molecules are strings of carbon atoms. But it could also have within it methods for changing some of its rules, as some programs already do. So, for example, it might have a rule which says: Every so often, if you can't solve a problem, try taking one of the constraints from that problem and considering the negative. When it's having a problem with benzene, it might say to itself, "I've got this constraint built into my thinking about chemistry which says that these molecules have strings of carbon atoms – in other words, they're not joined up. Well, let's consider that they *are* joined up." I don't see any reason why, in principle, you couldn't have such a system.'

As with other people I have interviewed, I asked her whether the kind of work she did diminished her sense of wonder at what it is to be human.

'Absolutely not,' she declared. 'AI has actually *increased* my sense of wonder at the human mind. What I mean by that is that all sorts of things we do and take for granted, things like speaking, or walking across the room and not bumping into chairs – things which, when people started doing AI, they assumed would be relatively straightforward to understand and therefore to program – turn out to be enormously subtle and complex. AI doesn't destroy our wonder; it's a way of deepening our wonder because it shows us a lot of unexpected richnesses which we hadn't realized were there.'

I invited her to imagine that she had created an artificial intelligence very much like a human being. Did she think that such an AI would start to wonder about its creator?

'Certainly it would wonder about its creator. It might adopt a sort of religious attitude. It might come up with the idea that there must be something out there which created the Universe and is responsible for the Universe. It might come up with the notion that this was (a) a very powerful and special transcendent person, and (b) a person with a particular interest in these complex computer systems. But I don't see any reason why it should (and what's more, if it did, I think it would be wrong). It might ask questions like: What am I here for? Is there a purpose to my being here? Yes, it would ask those questions.'

'Do you yourself think that life has a purpose?'

'It seems to me that, unless you start out with a religious notion of God the creator, and believe that we are carrying out some plan of his, I see no reason whatsoever to think that there is a purpose for human life. But I don't see it as necessary to believe that there is a purpose in order to live a human life with commitment and with passion and with enjoyment.'

'Is there not a problem with personhood coming into existence out of a cause which does not itself in some sense embody person-hood?' I persisted.

'Only in the sense in which there's a problem about striped things – black and orange striped things, like tigers – coming out of things which are neither black nor orange, still less striped, like atoms.'

'Except that striped things are clearly *physical* things. One can certainly understand how physical properties emerge from basic physical building-blocks like atoms. But if you are to deny person-hood to a First Cause, you are then committed eventually to an explanation of consciousness which is purely physical. It's just an emergent physical thing.'

'Well, it isn't a physical phenomenon. Take free will. I would say that it isn't a physical phenomenon. You can't describe in terms of atoms. It's a different level of description; it's a different level of order. Just as you can't describe the phenomenon of being stripy in terms of atoms; it's a question of the ordering of atoms. Stripiness and free, deliberative, careful choice are phenomena which exist at a higher level of organization. So you would not have to give up those concepts if you gave up the notion that there is a creator who is himself (itself) a person.'

*

I wondered how a theologian would react to all this. I put it to John Habgood: 'Suppose it were possible to create something as sophisticated as the human brain, and having done that, we were assured by this machine that it was self-conscious. And suppose we constructed this artificial intelligence to be free of original sin – all the genetically influenced behavioural "baggage" that *we* bring along with us from having been a product of evolution (innate aggression, for example) – could we create something which would be perfectly *good*? If so, would God love and approve of such an invention more than fallible human beings like us?'

'I think we need to be clear about what we mean by "original sin". Although we do carry some evolutionary baggage which one might get rid of in your perfect computer, in my understanding original sin is about the God-shaped blank in us – the absence of God from us.

'Now I suppose you could bring up your computer very religiously. Then why shouldn't such a marvellous computer love God and be loved by God? There's nothing so odd about making brains; we're doing it every day.'

I was puzzled. 'In what sense?' I asked.

'Every baby that's born. The fact that it's made of proteins and all that, rather than of silicon, may not make all that difference; not if, in the end, we think of a human person as a bundle of information – which, in a sense, is what we are.'

I asked Keith Ward to imagine that one day we created artificial intelligence. Would he see that as a case of us humans usurping the role of God? And what would be the *spiritual* status of such creatures?

'It's a little hard to say before we've done it, isn't it? But I don't see any reason why, in principle, it shouldn't be open to us to create intelligent and rational – and spiritual – beings. I would be prepared to baptize them.'

'You would?'

'Yes. I can't foresee it in my lifetime, but I wouldn't rule it out.'

'Suppose one were to program into them good behaviour: they

weren't to be selfish or sinful like we are, they were to live absolutely perfect lives. Do you think that God would value such created intelligences more highly than us fallible human beings?'

'There are two points here. One is that people like us are not very liable to create something which is morally perfect! I think we put our own defects into whatever we make. The second point, however, is that if we really managed to create one of these spiritual beings, it would of course, like us, have freedom – and if it had moral freedom, then we couldn't make *sure* that it was always better than we were. So in fact they would turn out to be just like us. I don't know that God would like them better, but,' he added with a laugh, 'they'd be less trouble.'

Chapter 4

Room for God?

Many of the people I spoke to talked about God. But what idea of 'God' did they have in mind? Was it any way affected by their knowledge of science? If the world is run according to the laws of nature, is there any room for God to have an effect – through miracles, for example? How does God fit in with the notion of time? In particular, does he know the future? And, looking to that future, what is to be *our* destiny – some kind of eternal life with God, or oblivion?

God in and beyond the laws of nature

We begin with the thorny problem of miracles, meaning events that run contrary to the normally accepted laws of nature. Given our modern understanding of science, what should be our attitude towards accounts of miracles?

Peter Atkins' response to my question was typically forthright: 'Oh, contempt! There are always reasons for disbelieving miracles. Most of them are connected with the reliability of the reporter. If somebody comes along and says they've seen the Virgin Mary at the top of the church tower, one should be very sceptical of that – just as one should be sceptical of people who claim to have seen flying saucers.

'David Hume, I think, got it right when he said that there is always more reason to disbelieve the reporter than there is to believe in what is reported: people wanting publicity; people having genuine hallucinations; a trick of the light; or whatever. If we did, in fact, discover that water was turned into wine, it would have a scientific explanation.

'Of course, we have to be careful about what we mean by "miracles". It could well be that the miracles in the Bible are just allegories that enable the writer to make a particular point for the audience that he had in mind – the particular uneducated audience that was likely to be listening to him.'

So as far as Peter Atkins is concerned, there is simply no possibility of miracles occurring. 'But,' I asked him, 'suppose you were faced with someone who is about to embark on a pilgrimage to Lourdes, and they're suffering, let's say, from an incurable disease of some kind. Would you discourage that person from going?'

He would. 'There have been examples of people going to this kind of place and dying as a result of it,' he declared. 'They thought they had been cured, came off their medication and dropped dead. So I think it's a highly dangerous game for people to ignore what is possibly the reliable way to good health – through science and medicine – and put themselves into the hands of priests, potions and prayers.

'Certainly, the workings of the body are so complex that one may be able to overcome some disabilities by will-power – by deciding that you're not going to hobble on your crutches any more, that you're really going to see if you can swing your legs. It may be that people require a sharp shock in order to pull themselves together. Going to Lourdes might provide the kind of shock needed. It seems an awfully expensive way of doing it. I think people ought just to say, "Right, I'm malingering, I'll stand up." And get on with the job.'

Richard Dawkins advances his own reason for dismissing miracles.

'One thing you could say about a miracle is that a scientist would judge it as very, very improbable. Many of us have had experiences like dreaming of somebody for the first time in many years and then waking up to discover that they are dead. These experiences are extremely disturbing. You think of it as a miracle, and it upsets you.

'We have to think statistically about this kind of thing. Every second of your life, some remarkable coincidence could be happening to you. In most seconds of your life it doesn't, and you don't report those occasions. Then, once in a blue moon, something does happen, and those are the ones that hit you. But you would expect that. You have so many seconds in your life that, during one or two of those seconds, you would expect to be experiencing some extreme coincidence – which you might think of as a miracle.

'Even less impressive are those cases where, for example, newspapers publish letters sent in by people who've had these experiences. You've simply got to look at the number of people who read the newspaper. If there are five million people reading it on any one day, quite a few of them are going to have astonishingly "miraculous" experiences. Editors wouldn't publish letters that say, "Dear Sir, I did not have a miraculous experience". It could be as simple as that.'

'Yes, but what about all the accounts of miracles in the Bible?' I objected. 'How do you think those stories got there?'

'It's easy to imagine that they got there by the same kind of process as *any* legend gets going in society,' he replied. 'Fifteen years after his death, there are numerous sightings of Elvis Presley. He has been seen all over the place! When a charismatic figure dies, legends abound.

'Sometimes works of fiction are mistakenly seen in later years as accounts of facts. Sometimes the culture in which stories were written down was one where people did not make the same rigorous distinction we do between a fact and a parable. It mightn't have been very important to them whether it actually happened or not. I've even met modern churchmen who, if I challenge them and say, "Do you actually believe in the Virgin Birth?", or "Do you actually believe in water changing into wine?", they say, "It doesn't matter whether it really happened. What's important is its symbolic significance." Sometimes they might just have been hallucinations. It's not a thing that I find difficult to account for.'

'What about miracles of healing?' I asked. 'One might have a take-it-or-leave-it attitude towards the miracles in the Bible, but today many people go to Lourdes seeking a cure from disease. Would you tell them it's a waste of time?'

'No, I would not. There is very strong medical evidence for placebo effects and psychosomatic illness. It's quite likely that if I persuaded this person not to go to Lourdes, he would go on having the disease, whereas if he went, he would be cured. *Psychologically* he believes he's going to be cured, and many illnesses (not all illnesses, but a significant number) are curable by psychological means. So I would feel quite irresponsible if I were instrumental in stopping somebody from possibly benefiting.'

'But if someone was suffering from, say, cancer, in your heart of hearts you wouldn't give much for their chances of being cured by going to Lourdes?'

'There are occasionally cancers which have been given up by the doctors and which do spontaneously regress. I wouldn't be so positive as to say that this *can't* come about by psychological means. But I'm pretty sceptical of the major kind of miracle cures where people suddenly get up out of their wheelchairs and start walking. In some cases, I suspect they were malingerers in the first place.'

Will Provine sees belief in miracles as an indispensable, but mistaken, feature of religion.

'Miracles are essential to a robust religious view of the world. If God merely started the world off in the beginning and then let it run on its own, or if God works through the laws of nature, what we have is an essentially atheistic world that just grinds on according to natural law. If you held either of those religious views of God, you might just as well be an atheist like myself.

'On the other hand, if there are miracles, it is a clear-cut sign that a supernatural force is working. So I think that religious people who believe in supernatural occurrences are saying just the right thing to give their religion some teeth and some substance.

'From a naturalist's point of view, I am extremely sceptical that miracles occur. However, I am open to demonstration. So if someone can take me to the lake where Cornell University is located, and have that lake parted (it's about 700 feet deep) so I can walk across, I would be enormously impressed. Indeed, that would turn me into a genuine believer in the supernatural.'

*

Theologian Nancey Murphy, whilst accepting that God usually works through the normal operation of the laws of nature, opens up the possibility of the occasional exception.

On my asking her whether she believed that miracles happen, she responded, 'Yes, I do. God has a large stake in maintaining the regularity and order of the cosmos. In fact, we couldn't even talk about a cosmos if there *weren't* regularities; it would be a chaotic mess. So most of the time God has to act in a very regular way – so that the world appears to be law-governed.

'But there are certainly occasions when God's purposes require some special sorts of act – especially if God is going to reveal God's purposes to the human race. So I believe that God not only *can* do that, but has done so on numerous occasions, and continues to do so in our daily lives.'

Did she think it worthwhile praying for somebody to get better, even if they have what people would regard as an incurable disease?

'Yes, I do,' she asserted. 'Extraordinary things do occasionally happen. I notice, though, that we most often pray for events where the outcome is inherently uncertain. In agricultural areas, people pray for good weather, and of course no one can predict the weather. We also tend to pray for people to recover from illnesses when we don't know exactly what the prognosis is going to be. So it seems to me that God most often works through the kinds of event that don't seem to be strictly ruled out by what we know of the laws of nature, but don't seem to be a sure bet either. God seems to prefer to act in such a way that it's possible to recognize his work if you are a person of faith, but where it's also always possible to *deny* God's action.'

'While believing that miracles can happen today, what do you say about the miracles we read of in the Bible? There are some very spectacular nature miracles going on there. Do you accept all those lock, stock and barrel?'

She thought for a moment before saying, 'I'm not sure yet. Certainly they are extravagant, and there's a good possibility that the stories were elaborated and exaggerated as they were passed along.

But there's also the possibility that if Jesus is who we claim he is (the Son of God – God acting in human life in a very dramatic and special way), it might well be that nature and everything around him would have behaved differently from what we ordinarily expect.'

I next asked theologian Keith Ward whether he believed the laws of nature could be violated.

'I do think miracles happen, but,' he added with distaste, 'I hate the phrase "violations of the laws of nature". It was invented by David Hume, who was a wonderful philosopher but a notorious atheist. And he invented the phrase to make miracles sound ridiculous: "Here's a law, so you ought to keep it; and here's a violation, so it's obviously bad." I don't like that way of talking about a miracle. I would go back to a more traditional view: a miracle is when physical objects transcend their normal powers of operation because they are transformed by the power of God. It puts it more positively.'

'But what would you say to somebody who has great difficulty in accepting, for example, that Jesus actually did walk on water, or actually did change water into wine? Is it important to believe in that sort of thing?'

'It's not important that those miracles happened in literal detail. Water into wine was primarily a symbolic story, about the new life that Jesus offered out of an apparently fossilized religious life. It doesn't much matter if a lot of miracle stories are symbolic, but it matters if you say no miracles ever happen. The reason is that you might be saying (or near to saying) God never does anything.

'Now God doesn't always have to do miracles, of course. But in the Christian faith the Resurrection is certainly pretty important. That has to count as an event which transcends normal human or physical powers. And if the Resurrection is important, you wouldn't expect it to be totally unique – just one funny thing happening in the whole history of the Universe and nothing else. It would have to fit into some pattern. You have to look for the *pattern:* the transformation of the physical by the spiritual. In other words, a sort of paranormal activity.

'People who don't like to talk about miracles often find no trouble at all in talking about the paranormal. I think most people have a suspicion that there are paranormal activities, and that they're connected with certain mental and spiritual powers. That seems to me to make sense. I would expect a person of great spiritual insight, who had a very close relationship to God, to be able to do things that most people aren't able to do – healing, for example. It's events centring around certain holy people or religious teachers which show the presence and power of a supreme spirit. In that sense I think miracles are quite important.'

Michael Poole is a visiting research fellow in science education at King's College, London, and has written on science and religion. I asked him whether his initial training as a physicist coloured his attitude towards miracles.

'I don't think it had a great deal of influence,' he told me. 'I became a Christian part-way through my physics degree, but the thinking out of these issues has really come since.

'I don't see a problem with miracles like the Resurrection – which, of course, is the central one of the Christian faith. Science, after all, is only investigating and trying to produce concise expressions for the *normal* patterns of behaviour. Miracles, of course, are not that – they are unique, one-off events. I like the phrase Charles Kingsley used when he described scientific laws as the "customs of God" – with the idea that, if God chooses to act differently on certain occasions for particular purposes, that's up to him. If God was capricious – if, say, gravity acted downwards today and upwards tomorrow – then of course life wouldn't be livable. But that doesn't mean to say that he *can't* act differently if he wants to for a particular purpose.'

'If God can intervene in these exceptional ways, on what basis does he decide? Do you see it as a kind of "afterthought" of his?'

'No, I don't think it's that. I'm not very happy with the word "intervene". The difference between the miracle and the everyday is not the fact of God working, but the mode of his operation. God is just as much at work in the natural, everyday things as in the unusual, one-off events. *Why* does he do it, and *when* does he do it? That, of course, presupposes some insight into the mind of God. But obviously, if we take the central miracle of the Christian faith – the Resurrection – then that was something worked towards over a long period of time: Christ should defeat death and come out the other side.'

'So although it was a one-off event, it was part of a plan?'

'Oh yes, absolutely,' he agreed, 'a key matter. Without it, as St Paul said, we should be of all men most miserable, because there wouldn't be any hope.'

Geneticist Sam Berry also believes in miracles. But did he believe *all* the miracles we read of in the Bible? 'I see no reason not to,' he said. 'We are told this is God in action. To me, there's no problem.'

'But,' I persisted, 'is it not a concern that miracles seem to have happened more often in the past than they do today?'

'There are places nowadays where miracles seem to be more common than in our developed Western world. You hear remarkable stories from the mission field about God apparently working in a most unexpected way. You can't tie God down, you can't say that he should be doing more miracles at a specific time or place. He works normally through normal mechanisms. He works through traditional medicine, traditional drugs. But there's no reason why he should not, for his own purposes, in his own time, carry out a specific healing. He may do it here, he may do it in Rwanda. He's more likely to do it there because they haven't got the benefit of the science that we have.'

'And I take it you're saying that what holds for Rwanda today also held in times past; they didn't have the medicines either. But how do you react specifically to the story of, say, the Virgin Birth of Christ?'

'If Christ is going to be fully divine, he must somehow be different. If God is going to bring his Son into the world, he could, as it were, have snapped his fingers and produced a full-blown infant. But Christ was fully human as well as fully God. So, we are told, he had a normal mother but a divine father. I have no problems whatsoever with the Virgin Birth. To me, this is a theological necessity, and I'm not going to screw up my science by trying to think of a mechanism by which God could have done it.'

By now, one might have got the impression that there exists a clean-cut divide between atheists who reject miracles and religious believers who accept them. But this is too neat.

Arthur Peacocke, theologian and former biologist: 'How God interacts with the world is, of course, one of the major questions for our times. There is no doubt that the scientific picture of the laws and regularities that govern interactions in the world alters the backcloth against which Christian believers, and indeed any theists, are bound to conceive of God interacting with the world.

'As a young graduate research scientist, I came back to belief in God because I was more and more impressed by the inbuilt *rationality* of the world that the sciences were uncovering. Every question you asked the world – every answer you got from directing questions to the world – always proved to be more intellectually exciting and intelligible than anything human rationality might have thought of beforehand.'

'Are you saying that because the world is intelligible, you think it's reasonable to assume that an Intelligence put it in place?' I asked.

'Exactly. God is the creator of a world which expresses a super-mind, a mind more intelligent, more rational than anything we could conceive. If God is supposed to interact with the world by interfering with these processes and stopping them, (a) I want some very good historical evidence that this has been the case, and (b) I want to know how this is consistent with the kind of being whose existence I am postulating in the first place – namely, a God who gives a rational, ordered creation which is creative and has the potentialities of being creative.'

'Are you rejecting miracles?'

'A miracle, in the sense of an act of God in the world which breaks the laws of nature, seems to me ruled out because it is inconsistent with the nature of God. A God who is a miracle-worker like that would not be a God worthy of worship.'

'So do I take it you reject all the accounts of miracles in the Bible? And if so, what are those stories actually saying? How did they get there? Are they attempts to deceive us, or are they something else?'

'Well, I think you first of all have to recognize that people in the Bible did not have this idea of the regular laws of nature. That is really a post-scientific phenomenon. They just saw *events*. And all events were directly due to God, whether they were regular events or irregular ones, whether they were surprising ones or non-surprising ones.

'Secondly, at least some of the writers of the New Testament don't talk about miracles as "breaking the laws of nature". They talk about a *miraculum*, a wonder – or the similar Greek word for a "sign". They are signs of God being specially present, or they are events in which God's presence is specially perceived. They weren't necessarily events which were breaking the laws of nature.

'Then there is a whole group of what people have popularly called "miracles" which do figure very largely in the New Testament, and which cannot be eliminated by historical scepticism: namely, the healing miracles of Jesus. Now at that time there were other healers going around in Palestine. We know so little about the relationship between the mind and physical health that we at least cannot be sceptical about the possibility of people being restored to health by the presence of a charismatic figure of great personal strength – a figure in whom they had faith. We see examples of this kind of thing happening today. So the healing miracles seem to me quite possible in the light of our present knowledge.'

'How about the Virgin Birth?' I asked. 'How do you think that account gets into the Bible?'

'At the time when those narratives were written, and indeed for centuries afterwards, the processes of human conception were not understood at all. It was thought that the woman was a mere receptacle, and that the man deposited, as it were, all that was necessary for the human being to be formed within her – and that was that. But we have discovered, over the last 150 years, that in every fertilized human ovum there are X and Y chromosomes. In order to be a male child, you must have an X chromosome, which comes from the mother, and a Y chromosome from the male. Y chromosomes only come from males. So what are we saying when we're talking about the virginal conception? Biologically, we are postulating that in Mary's ovum (which would only have X chromosomes) a Y chromosome was suddenly created by some miraculous waving of the wand. I say this because Matthew and Luke agree that Jesus had a nine-month gestation period: he started life as some kind of fertilized ovum. Where did the Y chromosome come from? If God is creating a Y chromosome in order to get the Jesus embryo to grow in Mary, what genetics did he put in there? Did he make sure that the nose Jesus had would at least look like Joseph's nose? One gets to a point where you're piling one absurdity on another. One's sense of plausibility breaks down.

'Combine this with the fact that the evidence, unlike the evidence for the Resurrection, is weak. The oldest New Testament narrative, St Mark, knows nothing of it. When St Paul is arguing about the significance of Jesus the Son of God, in the sense of the anointed human figure who conveys God to his contemporaries, he never refers to the Virgin Birth. So in a way, I would say I believe what St Paul believed. It was clearly no major part of his theology and outlook.

'Secondly, the historical evidence is extraordinarily thin. We have two narratives, Matthew and Luke, which cannot be reconciled with each other historically. Both have very strong legendary overtones. Few New Testament scholars regard them as historical narratives. They seem to be legends or stories told to emphasize the character and significance of Jesus; they are meant to convey a truth which is not historical.

'The Virgin Birth is a story developed to explain the extraordinary power and significance of the person of Jesus. Jesus was capable of expressing in himself God's own character of self-offering love. It was quite common, in ancient times, to attribute a special kind of birth to some outstanding person. I don't think we have to do that. If Jesus was to be fully human, he not only had to be bone of our bones, skin of our skin, but also DNA of our DNA. I think the idea of the virginal conception conveys a truth like the Adam and Eve stories – and that's how they should be regarded by modern Christians.'

A similar reluctance to accept miracles was shown by philosopher of religion David Pailin.

'I certainly don't find much happiness with the idea of God as intervening in specific ways, breaking into the situation. If that is the case, one begins to have great difficulties with the problem of evil. If God can stop a cell mutating into a cancerous form, why doesn't God do it? I think that's a very proper question. There doesn't seem to me to be any excuse for saying, "Well, there is a greater design, and it's all for good".

'If God came and made the divine will so absolutely clear that no one could possibly doubt it, it would cause grave problems for the idea that we have freedom. It would be rather like the situation when I am invigilating students taking an examination and stand behind one, reading what they are writing, and start muttering: "No, it's not that person; it's not Butler, it's Locke who said that." It would be a very brave student who would carry on writing "Butler" rather than "Locke".'

'So you see it as part of the scheme of things that God has to some extent to *hide* himself from us?'

'I think so. To be a hand's breadth off, to allow us to be free – but creating that gracious atmosphere that encourages us to be creative and to respond; providing that stimulating ambience, so that we do not settle down into dull uniformity but are urged on, and find satisfaction in being creative. That is the image of God I find easiest and most satisfying to work with.'

'So how do you react to the stories of miraculous events in the Bible? Are they an embarrassment to you?'

'I don't know if it's an embarrassment. They reflect the ways that people understood things in that culture. And it was a very different culture from the one in which we live now.'

*

John Habgood took up the theme of God being a God who for some reason chooses to 'hide' himself.

'God's problem, if I may put it like that, is to create a world which can be itself. If there is to be love which responds to God, God has to stand back in order to allow the world to be itself, to allow the love which flows back to him to be free.'

Arthur Peacocke has already said that he does not believe God interferes in the smooth running of the laws of nature. But if that is so, how did he see God making his presence felt?

'The God I worship does influence the world, but it seems to me influences it through *patterns* of events, none of which abrogate or break the regularities which God's own self has built into the system.

'Let me explain what I mean by this. When two people interact, all they do is send messages to each other by electromagnetic light waves, sound waves, pulses through the air, molecules, smell, physical touch. There is no other way in which persons communicate. And yet the most profound experiences any of us have are in personal relationships. What's going on here? Patterns of physical events are conveying meaning.

'Now it seems to me that God is doing this: communicating to the world all the time through various patterns of events, none of which break the laws of nature. I think God is not a person, but God has characteristics, some of which are like those of a person. And the most important feature of a person is that they communicate with each other. It seems to me God is a communicating God, and we must look for patterns in the world in which God has communicated. Perhaps there are patterns in history where God has expressed God. Perhaps there are patterns in human life. Christians believe that the pattern of the life of Jesus of Nazareth is a particular expression of God's meanings and intentions.

'God, it seems to me, can affect patterns of events by an input of something I can only describe as akin to what engineers call "information". Information can differ in its patterns without differing in its energy. In the end, I am driven to seeing God as interacting with the world by inputting information that affects the patterns of events, including patterns in our own brains. So perhaps God can influence our thoughts this way. But in none of these pattern formations is God actually breaking any of the regularities of the world whose existence is given to it by God's own self.'

Theologian Ian Barbour echoes some of the ideas of 'pattern' and 'information' expressed by Peacocke.

'I suppose the primary way of operating is that God has set up a system that expresses God's will. Evolution is God's way of creating. It impresses a lot of scientists that the conditions were just right for the Universe to expand at the right rate for life eventually to develop, and that molecular structures form in certain ways. Much of this has a kind of built-in purpose. One can talk about God sustaining the process and being the ultimate explanation. There is an element of chance, an element of creativity, in the process; the outcome isn't all laid out in detail in some divine plan that was preordained. There's flexibility in the process, and the unexpected. So that's how God continues to operate.

'Maybe there's something like an input of information. There are ways of talking about interaction which are not in terms of pushing things around or acting externally, but more in terms of the information content. Maybe God *informs* the world; maybe God expresses his way more from within. This is the Biblical idea of the Holy Spirit of God working from within, not externally manipulating things from without.'

Theologian Bob Russell firmly believes that God interacts with the world through miracle, but sees him at work in other ways as well.

'With most theologians, I would say that the very fact the Universe exists is an example of God's working. God is the creator, by which one means that without God there wouldn't be anything in the first place.

'Second, I think you can see God as working through time, sustaining the Universe in existence and guiding it. The laws of nature can be seen as an example of God's presence and activity; the regular workings of nature, as it follows those laws, can be seen as a form of God's faithfulness – God's covenant that nature will be dependable. As a free moral agent, I am able to act in the world to achieve intentional goals. (At least, I am convinced I am. If you are purely a reductionist, you would say I am not. But I don't buy that.) If *I'm* able to achieve goals in the world – without, of course, violating the laws of nature – I would imagine God *also* can in some ways achieve God's goals, not by violating the laws of nature but by working through them and transcending them.'

I asked him to elaborate on the idea of his being able to achieve his intentions even though the laws continue to run their natural course.

'I can choose to jump off a building or not,' he suggested. 'I'm convinced this is a free choice. Gravity doesn't determine whether I'm going to jump; it does determine that *if* I jump, I'll fall. So I can't violate gravity, but I can choose whether or not to fall.'

What emerged from these conversations was the conviction that, regardless of whether one accepts the occasional miracle, God can in any case be found expressing God's self even in the regular operation of the laws of nature.

God in and beyond time

We've heard conflicting views about the way God is supposed to relate to the laws of nature. How are we to think of God in relation to *time*?

Our modern understanding of time we owe to Albert Einstein. Before he formulated his Theory of Relativity, people regarded space and time as two quite distinct and unconnected entities. Indeed, that's the assumption that still underpins how most of us live our day-to-day life. But in my work as a professional physicist – when I do experiments involving extreme conditions of high speed – time and space reveal their true nature. Let me explain.

Suppose, for example, we have a mission controller at Houston, and directly above him a fast-moving astronaut flying in a spacecraft. At some distance away – in London, say – we have a schoolgirl demonstrating to her teacher that she can count: '1, 2, 3, 4, 5 ...' and so she continues.

As far as the mission controller is concerned, at precisely noon Houston time the London girl has got as far as reciting the number 3. Now the interesting thing is that, as far as the *astronaut* is concerned, at precisely noon Houston time the girl in London has not got to 3 – she's only got as far as saying '2'. And if he were flying in the *opposite* direction, he would conclude that the girl had not only said '3', she had by now got as far as '4'.

How do these differences come about? It's all to do with the fact that the mission controller and the astronaut are *moving relative* to each other. What is happening now in London according to one, is not what is happening now according to the other.

This raises a problem. We generally think that what exists is whatever is happening now, at this very instant. The past? That has ceased to exist. The future? That has yet to exist. So existence is exclusively a characteristic of whatever is happening right now. But as we've just seen, according to relativity theory people in relative motion don't agree as to what is happening right now! So if they can't agree about that, how can they agree about what exists?

The way one overcomes this problem in relativity theory is to say that EVERYTHING exists: The girl saying '2', the girl saying '3', the girl saying '4', and so on. All these occurrences exist on an equal footing. Past, present and future, there's no distinction; they all exist.

This, of course, is very hard to accept the first time one hears it. It is so counter-intuitive. Some people, no matter how often it is explained to them, remain unconvinced by the arguments, even when they are demonstrated in precise mathematical form. But as a physicist, I have no alternative. I *have* to accept it. It's the only way to account for our experimental results.

Now you may be wondering what relevance this excursion into modern physics could possibly have for our understanding of God. What intrigues me about it is whether it has anything to do with God's foreknowledge. A traditional understanding of God in relation to time is that God knows the future. This has never been an easy doctrine to accept. How can God know the future, it is argued, when I for one have not yet made up my mind what I'm going to do next? But what if, as we have just seen, the future in some sense does exist on an equal footing with the present?

I went to have a word with fellow physicist Chris Isham. Did he not think relativity had a bearing on the question of God knowing the future?

'I think it is relevant,' he agreed. 'It's an interesting question, how you conceive of the existence of time if you're a physicist. Clearly, our own human experiences are that tomorrow hasn't yet come and yesterday's gone. But the mathematics of physics is quite different. If you look at the mathematics of Einstein's theory, you find that you are presented – whether you like it or not – with a single thing called the "space-time continuum". There's then a nat-ural tendency to jump from this to saying that the whole of space and time therefore exist, in some way, as a single block entity.

'The people who work in process theology and philosophy are particularly anxious about this idea. There is a strong school of theological thought which says that God's knowledge changes, so God doesn't know what's going to happen tomorrow, as it were. And I must say I always found this very difficult to reconcile with the ideas of relativity. I think my own reaction to all of this is that even talking about God having "knowledge" as an ordinary human concept is very naïve and very simple. I would have to say that, in so far as God does "know" in the ordinary sense, he or she does indeed know the whole of reality, past, present and future.'

'Can I ask how you try to reconcile the apparently competing claims of yourself as a physicist, having to deal with all of time existing, and yourself as an ordinary human being, dealing with the fact that time appears to move relentlessly forward in a given direc-tion? Or can't you reconcile it?'

'I can see very clearly in the mathematics how reality (if I can call it that) looks as if it's timeless; it looks as if everything is somehow co-present. Yet I'm also, of course, conscious as a human being that that's not how we see things. I worry about it. I simply cannot reconcile it. I'm puzzled.'

I told him I was in the same boat. Did he have any hopes that eventually there will be some kind of reconciliation?

'If there is, it certainly won't be done just by physicists,' he ventured. 'It has to be that the notion of consciousness somehow plays a fundamental role. After all, it's our conscious experiences that move forward (or appear to move forward) from the past into the future. It's not just a job for physicists – it's far more than that.'

Former physicist Bob Russell regards the findings of relativity as definitely relevant to the question of God's foreknowledge.

'It's an important argument. Relativity says that there's no unique meaning to "the present". I can argue that God has knowledge of the future, without saying God calculates it *from* the present or foreknows it *through* the present. That would be inadequate to my experience of genuine openness, of freedom, of novelty, and of moral agency. But I think God knows every event because every event is within God's divine being. But in knowing that future event, God knows it to be my future – which is open to me from the present.'

Theologian Nancey Murphy does not see relativity as adding anything particularly new to our understanding of God in relation to time.

'I don't think that the changes in science made much difference to our view of God's foreknowledge. We have always claimed that God is in some sense outside of time or beyond time. Also, we understand that God gets his knowledge in a different way than we humans do. So while we're dependent on events in the present and whatever we can predict or project from those in order to know the future, God is in some sense present to the whole of time, however we describe that in physical terms.'

'Does it affect the relationship you have with God, the idea that he knows the future and you don't?'

'Yes, it does. For one thing, it means that I can have confidence in God. If God didn't know how things were going to come out in the end, then creating the Universe would be a huge crap-shoot. We have to justify a lot of the suffering and evil in this world on the assumption that God really does know how it's going to end up, and that the end is going to justify the process. So our relationship with God is based on that assumption.'

'Since God knows the future, he obviously lies somewhat out of time. Does that affect the sorts of things that you pray for, and when you might pray for them?'

'Yes, it does. This probably sounds a little silly, but on occasion I've prayed for events that are already passed – in cases where I don't know what the outcome was. If God is essentially outside of time, then God would have known, before that event happened, that I was *going to* pray. And so, while we think it's silly for us to try to affect events in the past, it doesn't seem so entirely out of the question to ask God to take care of events that are already passed in *our* time – because of God's availability to present, past, and future.'

'I wouldn't think that was at all silly!' I said. 'I do exactly the same thing myself. You're the first person I've heard of who does it too.'

Earlier, Chris Isham mentioned that some theologians don't like the idea of God having foreknowledge. Keith Ward is one of them.

'I actually disagree with most traditional Christian theologians on this one,' he told me. 'They tend to say that time is all there, from the first moment to the last, and that God knows it all. It's not really *fore*knowledge because God knows it outside time – he doesn't know it earlier in time. There's no problem with that; it's very straightforward. It just happens that I don't agree with it. In fact, I think a majority of theologians now disagree with it. We think that time has more reality than that. Theologians tend to say that time has a sort of successive reality because it allows you to be creative, to do new things. Now the price of that freedom and creativity and openness is that you can't predict the future exactly – not even God can. That's a non-traditional religious view. But it seems to many of us to be involved in what human responsibility really requires: freedom to determine human life. And indeed, *God's* freedom to do new things as well requires an openness of time.

'God doesn't know the future in detail, but God can still *control* the future if God wants to. God can certainly make sure it doesn't get out of control altogether. God can allow a lot of unpredictability, but can in fact be in control of the grand design.'

David Pailin has similar reservations about God knowing the future.

'I find the idea that in some sense the future already exists very puzzling,' he confessed. 'I want to know a lot more about what you mean by the phrase "in some sense". I have had my lunch; I have yet to have my dinner. I don't think anybody knows what my dinner is going to be until I've got around to having it. I don't think even God knows what the future is. I think the future is open for God as well as for all else.

'When you talk about God having foreknowledge, I am extremely suspicious. God knows all that there is to be known. God doesn't know the future because it isn't there to be known. There are *possibilities*; God will know what they are.'

'How do you account for statements like the one Jesus made when he said, "Before Abraham was, I am"? Doesn't that give an intimation that, as the Son of God, he had a rather special relationship with time?'

He was not convinced. 'I don't see that as dealing with time. I think it's maybe a claim about Jesus's closeness to the Divine.'

'Some people would say that if you read the Bible, you see God working his purpose out in history. It seems to be done in such a surefooted way that it could be intimating he knew what was going to happen. Wouldn't you say that?'

'I would not,' he retorted. 'If you look at events like Auschwitz, to hold that God has all this planned out, or that this is what God wants, is to me just like blasphemy. What I would want to hold is that whatever happens, God will be aware of it, and be able to respond to it, and seek to draw out of it what is for the best. But there isn't a firm blueprint already there. What would be the point of the whole creative process if it's all fixed? It would be like having a film where we knew that every shot was already there and that we're just going to play it through. I mean, the whole thing might be interesting the *first* time, but if God already knew the end, it really would be a futile exercise. I would rather see the creativity of God more on the model of that of the artist.'

'St Paul refers to God's foreknowledge. How did he arrive at that understanding? Are you saying that he got it wrong?'

'I think so.'

That rather took me aback. 'He got it wrong?'

'Yes, probably. People have for ages held that God knows what will happen in the future. It gives people a kind of confidence. But if you think about it, it also makes nonsense of the other religious claim that there is personal interrelationship between the human and the divine. If God knows what I'm going to do and that's already fixed, the idea of God then seeking to influence me towards more profitable ways of activity would be ruled out. It would be rather like me setting students an exam paper but already knowing exactly what they are going to do – I might as well give up trying to teach them.'

'I agree with you that as conscious human beings living out our lives, we have to accept that the future is open,' I said. 'It would be stupid for us to try to live as though we didn't have any decisions to make, everything having already been decided. That is certainly a characteristic of normal human life. But,' I continued, 'when you say that because it's impossible for *us* to know the future, it must also be impossible for God to know the future, could that not be a case of us cutting God down to our size – putting our own limitations on to him, making him in our own image?'

He thought about this for a moment, then replied: 'This is always a danger with theology. Anybody who tries to think about God has to walk along a very narrow ridge between making claims that they cannot justify and being absolutely naïve and stupid. But when I reflect on the reality of God, I think that if God is creative, "creative" means "bringing into being" and "letting be". And that means respecting a certain degree of autonomy and openness. Also,

I just do not understand how I can use active verbs about God like "choosing", "deciding", "creating", "responding", "loving" and so on if there is not some kind of significant temporal order within which God works. The difference in principle between the Divine and ourselves is that we know we work on a limited time-span. There was a time when we were not, and there will be a time when we will not be around to influence what happens. (Even if we write wills, we can't be sure that they are going to be carried out!) Whatever happens, God will be present and seeking to influence it. Nothing can take God out of being, and that is the big difference. That's why I would rather talk of God's relationship to time as being one of eternity. God was always there, and will always be there; and there never was a beginning, and there never will be an end, to the Divine reality.'

'You seem to think that God's foreknowledge would somehow prevent us having freedom of choice. But I can envisage a situation where, for example, if I know someone very well, like my wife, I can look to the future. I can look at the *Radio Times*, see what's on the television and the radio, and I can pretty well guess what she's going to want to watch or listen to, and what she's going to cook for dinner, and things of that kind – just because I know her very well. Now the fact that I can, with pretty good certainty, predict what she's going to do doesn't, of course, entail putting any restrictions on her. I'm just predicting what her free choice is going to be. I would have thought that, if God knew us far, far better and more thoroughly than I know my wife, why shouldn't he know exactly what I'm going to do in all situations that are going to arise?'

He smiled. 'Your language has slipped over from "I can predict" to "I know". That is the essential difference, because if we are people and we have freedom, we always have the possibility of doing *otherwise*. If the Divine/human relationship is to be thought of as one of friendship, it seems to me that friendship dies if both parties find each other utterly predictable.'

'So for you it's important that we sometimes take God by surprise?'

'I think God enjoys that. I think that's part of what God is looking for as a creator, real surprise. It's sometimes said that you can't tell a joke to God because he already knows the punch-line. Well, if that's the case, it must be very boring for God!' he laughed.

Similarly, Ian Barbour has difficulties with God in relation to time.

'I wouldn't claim that I know God well enough to know whether God knows the future. But I do think there are problems in the traditional claim. It seems to me that there are cases, both in science and in human experience, where one has to say that the future is open. In physics, it used to be a very mechanistic kind of world (the Newtonian world) where, according to rigid mechanical laws, everything follows from what was there before. But I don't think that's the picture you get from contemporary science. It's a more open and flexible process. Scientists make no claim to be able to tell you when a particular radioactive atom is going to disintegrate; it may be a second from now, it might be a thousand years from now. And no scientific theory can tell you when that particular atom is going to decay. It can give you probabilities, yes, but it can't predict. There is an element of chance, of unpredictability, in the whole scientific world: mutations in evolution; unpredictability in what's called "chaos theory". So from science it appears that the future is not determinate.

'I think also of human freedom: if it's to be real, that means it hasn't really been decided yet – not until *I* decide. And if that is so, I don't think even God could know. To me, it would not be real freedom if it had already been decided, or if the whole span of history is somehow laid out in the divine plan. I think God has planned and purposed a process that has a direction, and leads to the kinds of thing that are important. But it isn't an exact blueprint of all that's going to happen.'

Warming to his subject, he continued, 'God not only doesn't know everything, he can't control everything. The old picture, which was particularly prevalent in the Middle Ages, of a kind of absolute monarch who rules everything in the world doesn't go well with either science or with Biblical religion. To me, it's a God of love, a God who has limited himself, who has given the world a certain autonomy, given us freedom. It's a self-limited God; it's not something imposed on God from outside. It's a God who perhaps even suffers with the world in some kind of way. This to me is part of the message of the Cross. God is involved in human history, and is a vulnerable God. That's a God that I can worship more authentically. It's a God who empowers, rather than a God who exerts power over us. One of the themes of feminist theology is that we picture power in the male model: controlling power. Maybe there is another kind of power: the power to nurture, the power to empower, the power of love.'

I came away from these discussions with theologians rather nonplussed. It is very strange, to say the least, that in a situation where modern physics appears to have come up with powerful backing for the idea that the future in some sense exists, thereby making it easier to believe that God could have knowledge of that existent future, some professional theologians have difficulty in accepting the traditional position. Perhaps this is an area where a physicist finds it easier to believe than a theologian!

There is a second way in which our modern scientific understanding of time has a potential impact on how we view God. In his best-selling book *A Brief History of Time*, Stephen Hawking concludes with the question, 'What place then for a creator?' To understand why he said that, we must return once more to the subject of the Big Bang.

It is widely held that the instant of the Big Bang saw not only the origin of everything that we find in the Universe, but also the creation of space. The Big Bang did not take place at some well-defined point within an already existing space – like a terrorist bomb going off under a car in a particular street in a particular town. Before the Big Bang there was no space. It was the expansion of space itself, from nothing, that was responsible for the phenomenon of the expansion of the Universe. Even today, as we observe a distant galaxy receding from us, that galaxy is not speeding away from us *through* space. It is more helpful to think of it as moving because the space between us and it is still expanding in the after-math of the Big Bang. As it were, the galaxy is being borne along on a floodtide of continually expanding space.

That in itself is mind-stretching enough. But the plot thickens. We saw earlier something of the close connection between time and space. The relative motion of the astronaut and the mission controller affected their understanding of how far the girl in London had counted by the time it was noon in Houston. In other words, relative motion through space and time is able to change ideas about what is happening at any instant of time at different points in space. Einstein showed that the most convenient way of accounting for this was to treat time as though it were a fourth dimension, to be added to the three familiar dimensions of space: up/down; forwards/backwards; left/right. Einstein said, 'Henceforth we deal in a four-dimensional existence, not a three-dimensional existence evolving in time.'

In other words, we are to regard time as inseparably welded to space in a four-dimensional *space-time*. One cannot have space without time, nor can one have time without space.

For us, the importance of this is that, if the instant of the Big Bang saw the creation of space, it must also have marked the creation of time. There was no time before the Big Bang, because there was no 'before'.

Even stranger is Hawking's own idea that as one imagines going back in time to that first instant, time might even 'melt away'. There might not even have been a first instant. Whether Hawking's speculation is right or not, we are still left with a situation of there being no time beyond the Big Bang.

That of course raises a problem for anyone who has a mental picture of God the creator starting off on his own (having existed for all time), then at some point in time lighting the blue touch paper – and there's a Big Bang. Can one still think of God as the creator of the Universe if there was no time for him before there was a world? Hence Hawking's provocative question on whether there was a place for a creator.

*

Philosopher Keith Ward has no problems with this.

'Here again, St Augustine is the best Christian authority. He was quite clear that God isn't just in space-time – as though space-time were a sort of receptacle and God was either in it or out of it. God creates the whole of our space-time.

'Now there are a lot of complications here. But I think there's no difficulty about the fundamental thought that God is *beyond* space-time, that there can be things which are not confined to our space-time. There can, for example, be *other* space-times; there can be realities beyond space-time altogether. It's hard to imagine, of course, but it's quite a coherent thought that there could be a reality beyond space-time which sustains that space-time in being. Indeed, a very important part of basic religious perception is that from within the finite spaces and times of our lives, we can actually *sense* something beyond them – something which is also in them and sustaining them. That's the heart of the religious view, I think.'

Keith Ward referred there to St Augustine. I have always regarded Augustine as a truly remarkable thinker: fifteen hundred years before there was any talk of a Big Bang and of time being a fourth dimension, he had already worked out that there could have been no time before God had created the world. He did this by noting that time only has meaning and reality in the context of change: objects moving about and changing their positions with time. If the world had been created so that nothing ever changed – everything was static, no objects changed their positions – time would be an empty, meaningless concept. It would not refer to anything. Moreover, he argued, if there weren't any objects at all (the world and its contents not having yet been created), let alone objects that moved, then *a fortiori* there could be no time. Hence, he concluded, no time before God's act of creation. But unlike people today who are disturbed by Hawking's question, this was not an issue that worried St Augustine, any more than it worries modern-day theologians like Keith Ward.

God beyond death

It is said that the one certain thing about life is death. Whatever we might think about God in relation to time, we all know that for each of us time will run out one day. In fact, time will eventually run out for the human race as a whole, and possibly for the entire Universe. I asked astronomer Jocelyn Bell-Burnell to describe our ultimate destiny.

She looked rueful. 'None of the pictures is very cheerful. One involves the Universe continuing to expand for ever and ever. One involves the Universe continuing to expand for a while, and then the expansion halting and the whole thing contracting again. Everything falls in on top of itself, and it ends up with a Big Crunch. And there's a third picture that's sort of half-way between those two.

'The Big Crunch is clearly quite unhealthy for life. The Big Expansion picture is equally unhealthy, but in a different way. What's going on in the Universe at the moment is that nuclear reactions are taking place in stars like our Sun. The Sun is converting hydrogen to some of the heavier chemicals. In the very distant future, we shall reach the stage where old stars go out because they've run out of fuel in their centres, and new stars will not light because there's not enough of the hydrogen firelighter left. So the galaxies will go dark and life will go out.'

'What about life on Earth – how do we expect that's going to end?' I asked.

'Provided we don't pollute ourselves, or damage ourselves with nuclear weapons, then what will probably happen is that in about 5000 million years' time the Sun will begin to run out of fuel and will dim. But the problem is not so much the dimming as the expansion associated with it. The Sun will expand and engulf Mercury, the innermost planet. It will continue to expand and will engulf Venus, the next innermost planet. The planet after that is Earth. Whether the Sun expands so as to engulf Earth or not, we can't say – it's a bit marginal. But in terms of life here on Earth, what will happen is that the surface of the Sun (at a temperature of several thousand degrees) will come extremely close. The Earth will be charred; it will be a cinder.'

So we are faced with a future in which the Universe either grows cold and lifeless as it expands forever, or comes to a violent end in a Big Crunch. As for life on Earth, that will have been burnt up long before. How ought we to react to these grim-sounding prospects?

For those who believe in life beyond death, there is little to worry about. This life is to be regarded as but a preparation – 'a vale of soul-making' – for what God has in store for us. Once our spiritual beings have been shaped in the cut and thrust of life on Earth, they take on an existence of their own. The mould in which they were formed, the physical world, can be dispensed with, the indefinite prolongation of life in the Universe never having been God's intention. But that, of course, does depend on the hope of eternal life being well-founded.

I began my investigation of this topic by enquiring whether an acceptance of life beyond death has always been an integral part of religious belief. Biblical scholar Ernest Lucas: 'One has to say that, from the point of view of the Hebrew religion, it hasn't – not until quite late in the Old Testament. There is a sense of something beyond death: they call it *sheol*, but it seems to be virtually non-existent. The psalmist can cry out to God, as in Psalm 6: "Why do you let me go down into Sheol, because I can no longer praise you there?" So there was nothing much beyond death. That was one reason why they were so concerned about having children: to continue the family name, and their own name and reputation, after them.

'And yet there is another psalm, Number 16, where the psalmist can say that he's sure his relationship with God, being fundamental to his existence, will continue beyond death: "You will not give me up to Sheol. You will not let your faithful one see the pit." He doesn't know how this is going to happen, but he's sure this deep dependence on God – this loving relationship with God – is something not even death will be able to shatter. He can say, "In your

presence, there is fullness of joy at your right hand for evermore".
That's the only sort of grasp he's got of it. But he's sure that a deep
personal relationship with his creator cannot be ended by physical
death.'

Somewhat predictably, Will Provine will have none of this.

'I believe that when humans die, they are really *dead*,' he
declared emphatically.

'If it is as simple as that, how come so many people do have this
belief?' I asked.

'Because none of us likes to think of being really dead. We have
such a feeling of being alive we just don't want to be dead. I can
understand the feeling. I used to share it myself. But now I don't
particularly wish to live after I'm dead. Friends of mine who are
religious tell me that I'm wrong, and that I will live after I am dead.
If so, I will be in for an enormous surprise. If, however, I'm correct
and we don't live after we're dead, my religious friends are in for
no surprise at all. They'll just be dead.'

'It seems to me you're in a no-win situation!'

'That's true.'

We laughed. Knowing Will to be a rather militant atheist, I
found his response predictable.

I felt I was going to be on more comfortable ground with Jocelyn
Bell-Burnell, a committed Christian. However, I had something of
a surprise when I asked her if she believed in the life hereafter. She
thought for a while, then ventured, 'I'm not sure. I think a lot of
me would like to believe it. But that's not necessarily proper
grounds for believing it. What we want to believe can confuse our
objective thinking.'

It is always a problem disentangling the truth from what we wish
to be the truth (shades of Freud here again). But not to believe that
there is more than this earthly life has far-reaching repercussions
for the type of God we would be dealing with. I put it to her: 'We
believe in a God who is a God of justice, a God of love. But we are
then confronted by manifold injustices: people dying young, suf-

fering terribly. If this life is all there is, that God simply cannot be just, cannot be loving. So isn't it a rational argument that, if there are other grounds for believing that God is a God of love and justice, there *must* be something more than this life?'

She smiled. 'That's the "jam tomorrow" model, Russ.'

I shrugged. 'Yes,' I admitted.

'I think the Church has backed itself into a difficult corner on this one, and I speak as a churchgoer. It says that God is all-powerful and God is loving, and then has to explain why there's suffering in the world. Saying it will all come right in the next world, I don't think is good enough.'

'What would your explanation be?' I challenged.

'To doubt some of the initial assumptions.'

'That God isn't all-powerful or all good?'

'Or chooses not to intervene.'

'But why would he choose not to intervene?'

'Well, you get in the way of free will if you intervene. You're not allowing your people to be adult.'

'Yes,' I protested, 'but there are many people who suffer not because they *choose* to suffer. They have terrible lives for reasons which have nothing to do with their responsibility or their choice.'

'That's right, yes,' she agreed.

'So if there is no "jam tomorrow", as you put it, where is there justice for such people?'

'This is why I think the Church has backed itself into a difficult corner. All the arguments it produces to explain suffering aren't really very acceptable if you look at them hard. I begin to wonder whether we're right in saying that God is all-powerful and chooses to use the power, and whether God is loving. I just can't make those square.

'I'm very impressed by Emily Brontë's last lines: "Though Earth and man were gone, and suns and universes cease to be, and Thou wert left alone, every existence would exist in thee." Not amenable to scrutiny by a physicist, but plenty of meaning for somebody as a human being.'

*

David Pailin was equally cautious when I asked him whether he believed in an afterlife.

'Not in the sense of carrying on as I am now. We are finite beings who have a limited life. But this doesn't bother me. Everything we experience is also shared by God, and preserved in God. I am not terribly worried that the person I was twenty, thirty, forty years ago is no longer around. In fact, I find it something of a relief that the further back I go, that person is no longer with us! When I think of the future, I would hope – if I live say another twenty, thirty years – that it will not be the self I am now that will be continuing. So why should I worry about something carrying on? Why not say, "I'm preserved in God – God is treasuring it and cherishing it"?'

'So when all your history and experience and everything is absorbed into God, you don't see yourself as then having any sort of independent existence?'

'No. I find a lot of puzzles in that. If, say, in twenty years' time I develop Alzheimer's disease and survive another ten years in a very ill and confused state, would I then come to in that confused state? If I did, it would not be of any great benefit to anybody – even myself – to keep that thing going. If you then say, as people have said, "Ah, but you'll be resurrected in the prime of life, as you were when you were thirty-five", what about the last twenty-five years of my life? Is that just wiped off; have I no identity with it? I am afraid that, if that happens, I'd spend an awful long time wandering round saying, "Who am I?" I wouldn't recognize myself. If there's going to be personal continuity, then I think it creates enormous problems.'

John Cornwell's close involvement with the effects of drugs on the mind makes him see a very close connection between body and soul. Rather than think of a ghostly soul inhabiting a body, he speaks of an embodied soul. I asked him, 'Does this not pose problems as regards life after death, when this particular body goes into decay?'

'Yes, that is certainly true,' he replied. 'There are very few serious theologians in this university of Cambridge, and very few I would think in other universities (with notable exceptions), who believe in an immaterial soul as opposed to the body, the two in some way separate. The problem is that this creates great difficulties for understanding what happens in the afterlife. If we as a religion, as a set of beliefs, no longer have any firm conviction about the immateriality of the soul, and indeed the independence of the soul, what are we and where are we between our deaths and the day of the last triumph when "we receive our bodies back"? Very few theologians would now stress the idea of the afterlife in quite the same way as they used to. This, I think, is going to cause increasing difficulties as we struggle to understand the human soul in a new embodied way for the future.'

What kind of connection did psychologist Malcolm Jeeves envisage between one's soul and the body?

'Any belief in the afterlife for me would arise out of my Christian beliefs,' he said. 'I wouldn't try to answer it in terms of any so-called scientific evidence. My original coming into existence was in one sense the result of a straight biological process. Likewise, when I die my physical death will be the termination of a biological process. I also believe that, in a sense, my original coming into being depended upon God's gracious activity. Any hope that I have of continuing after I return to dust also depends upon God's gracious activity. What I'm saying is I don't believe that, tucked away somewhere within me, there is a little thing labelled "the soul" which has its own natural immortality. That is not what is taught in the scriptures. Our belief is in resurrection; God chooses to give us continuing existence.

'Now if you ask "What's it going to be like?", the response the apostle Paul gave to the Christians at Corinth was, "What a silly question!" He said the important thing is that, in some profound sense, we shall still be identifiable; we shall have what he figuratively called a "spiritual body" or a "glorified body". It doesn't seem to be a doctrine of us floating about as disembodied spirits, but rather that we shall in some sense be re-embodied. My hope of an afterlife depends upon my belief that Christ rose, and that he was the first-fruits of the resurrection.'

Astronomer John Houghton was in no doubts about there being life beyond death.

'I feel very convinced about it,' he affirmed.

'Would you like to hazard a guess as to what it might be like?'

'It clearly must centre on a relationship with God, because if it's to do with anything, it's to do with being with God; that's the message we get from Scripture, and indeed from Jesus. Jesus is the one bit of evidence we have, the one person who has been there and has told us something about it. It is to be a life with him and with God.

'It's also very clear that it has to have a strong link with the material creation, because we as Christians talk of resurrection, of something which is not just a disembodied spirit. We won't be just spirits without any form of communication or expression. We'll have bodies through which we can express ourselves and ...'

'Some kind of non-physical body, presumably?' I suggested.

'A spiritual body, but nevertheless with some relation to what's gone on down here. That's also very clear. The body Jesus had was one which looked like an ordinary body at times, although it could appear and disappear at will. I take those appearances seriously, because they seem to convey the sort of message about Heaven which is probably important for us to receive – even though we couldn't conceivably understand what it would really be like.'

As we have seen from the last two contributions, much of a Christian's belief in an afterlife stems from the resurrection of Jesus. You will remember how Arthur Peacocke was very sceptical about miracles. I asked him, 'How about the Resurrection of Jesus, the empty tomb? Do you put that down to just another miracle story which you would discount?'

'This is a very interesting one. Science can comment on events which are regularly repeatable, in which event A goes to event B and to event C. In the case of the Resurrection, you have an end state – namely the resurrected Jesus – which is not within the purview of science at all. Clearly, it isn't a repeatable experiment. The only science it might come within the purview of is psychology. Can a group of people be convinced psychologically of the presence of a living person over a period of several days and in different circumstances? Can there be a kind of complex hallucination?

'My judgement from reading the narratives is that the people do not show signs of that. The Resurrection comes as a surprise; they don't expect it. They don't necessarily *want* it. But the fact that the total person of Jesus is somehow taken through death is part of the primary data that launched the phenomenon of the Christian Church.

'Was the tomb empty? I am agnostic about this. It seems to me that the statement of the Resurrection is not about an empty tomb. It never has been. St Paul doesn't mention an empty tomb; he talks about the experience of the resurrected Jesus.'

Granted that Arthur accepted the Resurrection of Jesus, did he believe in ourselves having a life after death?

'"Life after death" is a misleading phrase. "After" is a word that refers to time. When you talk about life after death, it isn't "life" as we know it; it isn't "after" as we know it. When our bodies dissolve, they will no longer be biologically viable. All that makes our brains active and helps us to think will dissolve. On the basis of science or what we know about the relation of mind and body, there seems to me no reason to support any belief that we as persons will survive the dissolution of our bodies as such.

'Christians affirm resurrection. By "resurrection" they are referring to what they believe happened to Jesus – namely that, by an act of God, the essentials of his personhood were taken through death and given a new kind of embodiment. It can't be a body in the form we usually mean, but some kind of embodiment in which the individual personhood can still be identified and still known to others.

'We have today a rather useful analogy for this. We are all familiar with software and hardware. We know that software programs can be embodied in many different kinds of hardware, and yet the software is real enough. The software which constitutes the pattern in our brains constitutes the essential you and me. It's at least conceivable that the essential information and pattern which constitutes our personhood could be given a new embodiment. Death is the dissolution of the present location of our personhood. But it seems to me possible that our personhood could be located in another form, in another mode of existence where God is.

'Now you might say, why should I believe in this? I don't believe in it because of any science. I believe in it because it happened to Jesus. Could it happen to us? I say, "Yes, it could happen to us because it happened to Jesus". *Will* it happen to us? The answer is, "Not necessarily". What is resurrectable is the kind of life that Jesus had – namely the life which is self-offering love devoted to God and humanity. That's what God takes through death.'

Like Arthur Peacocke, Nancey Murphy has a firm belief in resurrection.

'There are competing views of what life after death is supposed to be like,' she told me. 'Perhaps the most common view among Christians is that we have immortal souls which can leave the body at the point of death. But the Christian view is that the entire person is resurrected at some point, and we thus have an embodied existence; not the usual physical body subject to the laws of nature, but nonetheless a body that allows us to interact with one another. It makes community possible. The afterlife has to involve our living with one another according to the pattern that Jesus set up with his disciples.'

'So you would see the afterlife as involving time. There would be a progression of events; it's not just a state of bliss where nothing happens?'

'Yes, I think so. The idea that everything would cease to happen and become perfectly still and motionless is more a Greek notion than a Christian view.'

'Don't you worry about it getting boring doing the same thing for ever?'

'It's difficult, of course, to think of what one is going to do with all of that time. But I can certainly imagine using up a great deal of it! We actually need a lot of time to complete the projects we have begun in this life – to make friends with all of the people that are going be there, and so forth.'

I asked theologian Bob Russell whether he thought the passing of time had a role to play in the afterlife. What he told me takes us back to what we were saying earlier about the physicist's enriched understanding of the possibilities of time.

'I think the passing of time does apply. The two standard views of eternity are either of unending time (unending pizzas day after day, more and more of the same) or of timelessness ("All freeze!"). Both of those get at pieces of our intuition, but they are both inadequate. God's time, which we call "eternity", is more complex.

'For example, for us the past fades and the future is anticipated. I think the fading of the past won't be part of eternity, but the distinction between past, present and future, and the experience of moving from the past through the present to the future, will be.

'It will be in a sense timeless, because it won't have the aspect of mourning the past because it's gone and irretrievable (as it is for us now) or of never knowing what the future will be. It won't have that aspect because that's the *brokenness* of time, of our time now. What's exciting is that physics helps us here in a very profound way.'

'In what way?'

'Well, physics has opened up the whole question of time. We really live in a four-dimensional world, a space-time continuum. So just as I occupy a volume – perhaps too much of a volume! – so I also have a history. And that whole space-time domain is in some profound way *me*. Physics raises a serious challenge to our naïve view of what's past, present and future. I think the debate is just beginning.'

It's all very well entertaining thoughts of eternal life, of going to Heaven. But there is the other side of the coin. I asked John Houghton how he viewed Hell.

'Well, Hell is something you don't like thinking about because it's extremely unpleasant.'

'That's why I'm asking you!'

'I think the clearest view I have is of *absence from God*. In Heaven we are there in God's presence; that must be very aweful and wonderful and tremendous. Absence from God sounds absolutely dreadful; that must be what Hell is. If people don't want God, then the message is they can have that.'

Arthur Peacocke echoed the same idea. 'Hell is the absence of God. People are in Hell if they live in a world entirely enclosed by their own egos. Hell obviously isn't a physical location. It is a state of personal being.

'Hell is well depicted in that famous book by Dante. When you have gone through fires, the bottom of Hell is a lake of ice. Why a lake of ice? Because that is a symbol of where all sensitivities, all sensibilities, all potentialities, all possibilities are frozen out.'

I left the last word on Hell to Nancey Murphy.

'I grew up Catholic and we believed that there was Heaven, Hell and Purgatory. I've wished that when the Reformation came along, they'd given up on Hell and kept Purgatory, rather than the other way round! Hell is a difficult issue. On the one hand, it's hard to imagine that God would create any of us destined for eternal punishment. On the other hand, we have a great stake in God respecting our freedom. And so we have to keep an opening, I think, for people who will choose perpetually to reject God. But it's also important to note that the images of Hell in the New Testament are metaphors intended to give us a sense of the severity of the loss for one who is lost eternally. I think it's much more likely that we simply cease to exist if we choose to reject God in an irrevocable way.'

As we saw earlier, one of the features of relativity theory was that it introduced the idea of time as a fourth dimension. It took a long time for this to come to our attention because of course we experience that fourth dimension in a totally different way from the other three; it was not easy to make the connection. Could this lead to the possibility of there being still more dimensions: dimensions experienced in yet further ways, or perhaps not experienced at all (in this life at least)? Could those extra dimensions provide an answer to the perennial problem of where Heaven and Hell might be located? I tackled physicist Paul Davies about this.

'There are problems about adding more dimensions of time. These are technical problems having to do with quantum physics and the various forces of nature,' he explained. 'But extra dimensions of space can easily be accommodated. Of course, we have to account for the fact that we don't see them. There's a very natural and easy way to get rid of additional space dimensions by rolling them up to a very small size.

'Let me give you an analogy. If you see a hosepipe from a distance, it looks like a wiggly line. But when you get very close, you see that what you thought was a point on the line is actually a little circle going around the tube. In the same way, what we think of as a point in space might, if we could see it in fine enough detail, be a little circle going around a fourth space dimension. And of course, if you can hide one dimension that way, you can hide any number. There are theories popular at the moment (with theoretical physicists) which incorporate higher numbers of dimensions: ten space-time dimensions, or twenty-six, are numbers that endear themselves to theorists for a variety of mathematical reasons. At this particular time we simply don't know, but it's clearly possible to have extra dimensions.'

'Many people wonder where Heaven is. Do you think there is any mileage in the idea of locating Heaven in one of these rolled-up spatial dimensions?' I suggested.

'It's important to realize that the rolled-up dimensions are not somewhere else. They're *inside us*, so to speak, all around us. Every point in space, if we could scrutinize it through a fine enough microscope, would be seen to be either a little circle or a higher number of dimensions – if the theory is correct. When people have an image of Heaven as somewhere totally outside of, or different from, the physical Universe, I don't think you can accommodate that in something which is very much part and parcel of the physical Universe. I've never liked the idea of Heaven as just being somewhere else in the Universe, somewhere we can't see. It's like saying that when you die, you go to some other planet that we haven't happened to spot so far. It seems to me that if we're to take the concept of survival after death seriously, it has to be something much more profound than a sort of reincarnation in some other bit of the physical Universe.'

If we are not to take extra dimensions as providing a literal answer to the question of 'where' Heaven might be, there might still be some advantage in thinking of them as a useful analogy. John Houghton likes to put forward an analogy involving multi-dimensions to illustrate how we must be prepared to widen the scope of our thinking when confronted by questions about the nature and working of God.

'John, you have written in one of your books about "Flatland". You use this as an analogy to try to get across some ideas of spirituality. Could you explain what that was about?'

'Yes. Imagine a world in which we as beings don't have an up and down; we just have a north, south, east and west. So we're flat creatures living on a plane. We have to slide around on that plane.

'There was a fascinating book called *Flatland,* written a hundred years ago by the mathematician E. A. Abbott, which describes what happens in this world of Flatland. The punch-line of the book comes when a sphere from Spaceland comes into Flatland, and moves up and down through this plane. At first it appears like a point (as the sphere just touches the plane), then as a circle, and then as a larger circle, before it disappears out the other side. The Flatland inhabitants can't make head nor tail of this, because they don't understand what it means to be a sphere. Eventually, the sphere picks up one of these Flatlanders and takes him outside. They fly over Flatland and look inside the houses ...'

'Which if the doors of the houses were closed, and the windows shuttered, would be impossible while you're still confined to the plane.'

'Of course, quite impossible. So the Flatlander now believes in the existence of this third dimension. He goes back to try to explain to his fellow Flatlanders what it's like, and he's put in jail because he's thought to have gone bonkers!

'I think this is a very interesting picture for us, because we are in a world of perhaps four dimensions: space and time.'

'So, just as for the Flatlander the third dimension makes seemingly impossible things happen if he is thinking solely in terms of a two-dimensional plane, so the introduction of Einstein's fourth dimension also makes things which seem impossible ...'

'... to be possible. That's right. Now, we're in this world of four dimensions and we imagine that's all there is.'

'But we might be wrong again.'

'Yes. If you think of God as being in another dimension (just as an analogy, of course – I'm not suggesting he really is), that does help to stretch our minds about where God is and how he might influence the world. It puts God outside our creation in a fifth dimension, but also means he can be *within* it also; he can express himself within it.'

'And yet again, impossible things could happen?'

'Of course, yes.'

'Suppose we pursued this analogy of the fifth dimension. Can one get any more out of it?' I asked.

'Well, a world of three dimensions is much more *solid* than a world of two. A cube is more solid than a piece of paper. You could argue, therefore, by analogy that a world of five dimensions is more "solid" than a world of four dimensions. So because Heaven (where God is) has a five-dimensional structure (on this analogy), it could be much more solid, much more *real*, than a world of four dimensions. Perhaps it's the world of four dimensions which is a bit phantom-like.'

'You mean, in the same way as a two-dimensional shadow is not a particularly good representation of an object that in reality is a three-dimensional solid?'

'Very interesting ideas can come out of analogies of this kind, yes.'

Chapter 5

Science and Religion

Having sampled a range of reactions to a variety of scientific topics, it is time to draw the threads of our discussion together. In general terms, how should we see religion and science, and the relationship between the two?

Science without religion

We begin with the views of those who are entirely dismissive of religion. Richard Dawkins is well-known for his antagonism. In some of his writings he has attacked religion by likening it to a computer virus. I invited him to elaborate.

'Computer viruses are like real DNA viruses,' he explained. 'They're written instructions which flourish because they find themselves in a world (respectively computers and cells) where they are copied, and where they are obeyed. It's a very easy thing for a parasitic code to arise which simply says "copy me". This works in computers because the computer obviously obeys instructions. Computer viruses will be passed around in the form of floppy disks, or over the Internet, or something like that. When they're obeyed, they cause several copies of themselves to be made. They often have some destructive side-effect, although that's not important – that's just if the author of the computer program is mischievous. The point is that "copy me" instructions get copied because that's what they say, and they don't need any other rationale.

'Now DNA viruses are familiar: they cause colds and flu and other things. Computer viruses are familiar – we all know about them. The question is: is there any other medium, analogous to the cell and the computer, which is vulnerable to this kind of parasitic code?

'Human brains are a possible candidate. They are in constant communication with one another via language. A chain letter is an example I've used. These are where you get through the post something that says, "Please copy out what it says on this postcard six times, and send it off to six friends". That in itself is a kind of virus; it will spread exponentially. People ordinarily *won't* spread it. Why should they? It's a waste of their time. It has to be backed up by some kind of threat, or promise of reward. So they're told something like, "If you don't do this, you'll die in horrible agony." Enough people believe that to send the letter off, so that it does partially spread. Chain letters spread just like a measles epidemic, and then die away.

'Religion as a virus is a more contentious suggestion. It might work something like this. When children are young, it's important (for good Darwinian reasons) that they should listen to what they're told and believe it. And it's genuinely good that this should happen. A child is in the phase of life when it's sucking in a lot of information, recording it, obeying it, before later passing it on to the next generation. That is just the kind of climate where *parasitic* information could come in and exploit it. Cells are not designed to carry viruses, but if parasitic DNA arises, it will spread. So child brains, being designedly gullible, mean that parasitic information can spread. It would take the form, "Believe X, and then when you grow up make sure your children believe X as well." In principle, that will work whatever X is.

'I think that may be what religions are. Just as the chain letter is helped along by the threat or the reward, with religion it's not just "copy me", but "copy me, or if you don't you'll go to Hell", or "copy me, and if you do you'll go to Heaven".'

'This idea of yours that religion is a kind of computer virus infecting the system is a very powerful analogy. But can't it equally be applied the other way?' I protested. 'For example, when one looks back over the course of human history, one finds that just about every tribe, every people that has ever lived, were religious. You can tell that from the way they buried their dead. The phenomenon of atheism is rather modern; it's something which has sprung up in the industrial West. So couldn't one just as easily – and perhaps even more appropriately – describe atheism as a virus which is infecting the normal system?'

'Yes, there is a real sense in which to be religious is the normal system. It undoubtedly is the most widespread thing,' he admitted. 'But I don't want to use the word "virus" universally for any idea. I want to use it only for ideas that spread because they embody a self-copying instruction in themselves.'

'Yes, but I heard a series of lectures you gave at the Royal Institution (the Christmas Lectures) about natural selection. You were getting your atheism in as often as you could, and you were addressing a young audience. So I would have thought that you were doing as good a job of infecting gullible minds as any priest has ever done!'

He smiled. 'That's true. I was talking to children, but I don't pull my punches whoever I'm talking to. If we go back to the computer virus analogy, one could say that all computer programs are viruses. Good programs tend to spread around the world. In the days before companies worried about copyright, they spread even more! Computer games tend to spread because they are fun to play. They are all virus-*like*, because they are spreading. But I'd really rather say that they are analogous to proper genes which are doing good things, like making you run fast or see clearly. The thing about *viruses* is that they're spread by simply saying "spread me"; they don't say, "Do something good for the animal." The question is: are religious ideas in that category? Are the sorts of ideas I was telling the children in the Christmas Lectures in that category? The mere fact that they spread to children is not evidence that they're virus-like. To be virus-like, they've got to spread just because they are tailored to spread. If the children I told about evolution came away believing in it, I'd like to think that they did so because the ideas are plausible, or because they're backed up by evidence. If I'd taken any *religious* ideas, and any of them had believed those, it would be simply because I'd told them; it wouldn't be because I'd provided any evidence for them.'

Here we reached an impasse. With Richard unprepared to accept as evidence anything that could not be verified scientifically, and being of the opinion that religion was not good for you, we had simply to agree to differ.

One of the things about Richard that has always provoked my curiosity is what drives him. Why is he so militantly atheist, whether it is in his writing, his lectures or his broadcasts? I put it to him: 'If I didn't believe in God, and I was confronted by people who did, I would think, "Well, they're wasting their time. They could have spent Sunday morning in bed, instead of going to church." I would be rather indifferent if they were wasting their time like that. But you don't seem to be like that. You seem to be driven to fight almost as religiously against religious belief as I would be on the other side of the fence.'

'There is a lot in that. One of my reasons is that quite a lot of evil is done in the name of religion. I don't want to particularize, but historically it's true. I think it stems from the fact that religious belief is held with enormous conviction but without backing evidence. It is possible for two people to be absolutely convinced that they are right, and for there to be simply no formula for them to settle their differences by argument. Therefore the only way to settle it is by violence.

'But the reason why I'm driven is probably more to do with the fact that I have such a strong feeling for the Universe, and for how remarkable it is that we can understand how it works and understand life (to the extent that we do). It's such a richly rewarding experience to share in that understanding that I feel desperately sad that people are deprived of that experience by being fed what I see as inadequate – medieval, in many cases – substitutes for it.'

'If you're talking about those religious people who, for example, accept the story of Adam and Eve as being literally the way human beings came into existence (and therefore reject evolution), I understand how you feel,' I conceded. 'I go along with you completely.

'But in my experience, the vast majority of the religious people I come into contact with fully accept scientific explanations – including Big Bang cosmology and the theory of evolution. I feel just as driven as you do to engage the public in understanding my science. I think it's enormously enriching for them to understand something about modern physics and, in your case, evolution. I don't see any incompatibility between being driven to make the public more scientifically literate and being religious. Why should your drive to get across scientific knowledge also be coupled with such a strong drive to knock all forms of religion?'

'Yes, your explanations of physics (which I've read and greatly admire, by the way) seem to me a testimony to what you've just said, and I agree with that. But I think that people like you are very unusual as religious believers. There is a minority of people sophisticated enough to understand science, to believe in evolution and so on, and to couple it with their belief in God. But there is, I think, a great divide between the sophisticated theologians on the one hand and the people in the pew on the other. I don't think that churchmen exactly go out of their way to disabuse their flock. They're happy to let them go along with this unsophisticated view, although they don't hold it themselves. I think of it as an impoverished view. I know *your* view is not impoverished, because I've read your books. But most people who are influenced by religion are, I believe, impoverished rather than enriched by it. Without it, they would have a vision of the Universe, of life, and of their place in it which is bigger, more dignified, more uplifting.'

'Richard, there's a great deal there that you and I can agree about!' I concluded.

When it comes to the expression of anti-religious views, the name most readily linked with Richard's is that of Peter Atkins. The first time I heard Peter speak was at St George's House, Windsor Castle. A group of scientists, theologians and psychologists had been gathered together, at the request of Prince Philip, to discuss with him the relationship between science and religion. Peter was the first speaker. I reminded him of that occasion.

'I didn't know you at that stage, Peter, so I didn't quite know what to expect. You gave a talk about science, and I noticed that every time you used the word "science" one could have substituted the word "God". You talked about its awesome nature, and how it was all-powerful and magnificent; I forget your actual words, but it was along those lines. I was under the impression that it was a Devil's Advocate talk, that you were portraying science as God. But in the end, it turned out that you were actually being serious. Do you in fact see science as your God? Is it the thing that you believe in most, that you live for? What actually drives you?'

'I'm driven by the desire to understand. The world around us is extraordinary: it's delightful, it's wonderful, it's awesome,' he enthused. 'Science enables one to pick it apart, to look inside and see why it is so wonderful. I take a deep pleasure from understanding the workings of the world. So although you could in some places substitute the word "God" for "science" in what I say, there are many more places where you couldn't. You couldn't, for example, substitute the word "God" when you're talking about science as being a public activity, where people are doing experiments under controlled conditions and sharing their results and making sure that everything really hangs together. You couldn't use the word "God" in the context of science making predictions, or when science is being useful in developing new pharmaceuticals and building aeroplanes. I think that shows science has progressed beyond simple religion and faith in the unknowable.

'My motivation is really to say, "Look! This world is wonderful!" instead of just being told by written authority that it is wonderful because God made it so. Let's see whether we can use the extraordinary power of the human intellect to unravel it and see how it works. If we can see how it works, that does not diminish our delight. It adds to it. We still have the awe and wonder of what we see around us, but we also have the deeper joy of knowing how it hangs together and where it came from.'

'I share with you, to a very large extent, your love of science and the world, and your sense of awe and wonder. But on top of that, I also believe in God. You on the other hand have, on top of your sense of awe and wonder at science, a very strong atheism: one that you push as hard as I would push religion. I find that very curious. Why are you so strongly against religious belief?'

'I think religion kills. And where it doesn't kill, it stifles. Religion scorns the human intellect by saying that the human brain is simply too puny to understand. In contrast, science enables one to liberate oneself; it liberates the aspirations of humanity. Science is the apotheosis of the Renaissance; it is the flowering of humanity. It says that here inside the collective head of the human race is an organ of sufficient power really to understand what's going on. That ability to understand is so extraordinary, so precious, that it ought to be fanned into activity wherever it is found. The human intellect is one of the most precious things in the Universe.'

'You talk of science as being liberating and religion as being stifling. As a practising scientist myself,' I objected, 'I do not in any way see my religion as stifling my sense of curiosity. I am just as curious about the world as any other scientist. I would say that if you think only in scientific terms, surely *that* is stifling. It's restricting your thinking to only those questions that science is able to answer. Whereas religion, I would say, is liberating in that it encompasses everything that science can say about nature, but points to other realities as well, and other questions – questions to do with meaning and purpose.'

'Well, science can answer all those questions. I don't think there is any question that science cannot tackle. And I think that, as it tackles them, it gives people answers that are much more reliable, much more plausible, than the obscure arguments religion provides. I mean, many of the questions religion tries to answer are not real questions. Take one that you've just mentioned: the purpose of the world. In my view, that's an entirely invented question. People sat around on dark nights in the rain, wondering why they were here. Why do we have to go out and kill mammoths every day? Why do you just sort of slave away? There must be a purpose. And religious institutions – churches and so on – have fanned the notion of purpose into a deep question. But in fact, it's a totally empty question. So science gets rid of it by saying, "Look, you've invented it."'

'Science cannot answer the question, "What is the purpose of life?" and therefore dismisses the question. That, I would say, is a perfect example of how stifling it is to treat science as your God.'

He thought for a moment, then suddenly asked, 'Why is there a teapot in orbit around Mars?' He looked to me for an answer. 'We could spend two thousand years worrying about why there's a teapot in orbit around Mars,' he continued. 'What science says is, "Come on, there *isn't* a teapot in orbit around Mars. You know that jolly well – you've just invented it. It's an entirely human invention." The problem of why there's a teapot in orbit around Mars is identical (a bit more risible, but identical) to the question, "What is the purpose of the Universe"? Science can show that there is not a purpose in the Universe, and is not going to waste its time worrying about it.'

By this stage of our conversation, I felt that Peter had again demonstrated very clearly what I had asserted at the outset – that science was his 'God'. My mind went back to what some Jungian psychologists had told me over breakfast at St George's House the morning after Peter had delivered his talk there. They explained that in the psychological theory of Jung, the religious drive is so important and all-pervasive that if it does not exercise itself in the worship of God, it will devote its energies to some substitute for God. Their view was that Peter's performance the evening before was ample verification of that aspect of Jungian thought!

Yet I persisted. I demanded of him: 'Do you see any limitations at all to science? Do you look forward to a time when science will be able to answer *all* the questions you regard as meaningful?'

'Oh yes, certainly,' he declared. 'There's every reason to suppose that science will come to an end in understanding the fundamentals of the world, the composition of the Universe, the origin of the Universe – the "what makes it tick" questions. But we have to be very careful of talking about the "end" of science. In a sense, science is like an infinite length of string with just one end. The string being infinite means that the future of science is unbounded: there will be no end to understanding the complexities that can arise from this understood universe; there will be no end to technology, that would be another way of putting it. One should be very cautious about what it means to have *complete* understanding.'

I left Peter's study at Lincoln College thinking that the science and religion debate would certainly be a lot tamer were it not for his spirited contributions. Though in scientific circles he is probably something of a one-off, in that few professional scientists would express their devotion to science in quite such extreme terms, Peter does nevertheless voice the views of a sizeable proportion of the public at large.

The majority of scientists on the other hand, atheist or otherwise, seem to sense that, powerful and valuable though their science might be, it does not have all the answers. They accept that just because science cannot come up with an answer, it does not mean that the question has to be ruled out of court.

Will Provine, for example, told me of his concern that some of his students are inclined to draw wrong conclusions from their science when it comes to morals.

'There's a growing number of young people who believe pretty much as I do, that they're naturalists in looking at the world. But they then conclude that morally anything goes, because there's no ultimate morality in the world. I think this is very dangerous and very sad. It tends to destroy the proximate meaning in life that young people could have for themselves if they were brought up to understand more about the interactions of human beings.

'So I go on my personal crusade, which is to try and reach the students in a university setting who are already naturalists, but who have not thought very much about how to find meaning in life, nor very much about how to get along with other human beings. It's terribly important that religious people begin thinking about the foundations of ethics and meaning in life for *non-believers*, as well as for themselves.'

I asked self-professed materialist Steven Rose whether he saw human life as having any kind of purpose.

'I find that my own life has purposes,' he said, 'and they are those that go with my science; that is, trying to understand the world, and also trying to change it. There's an incredible account in today's newspapers of the World Health Organisation's report on how a large mass of the world's population is living and dying in poverty. I regard that as an outrage, and an outrage which is changeable. I want to create a society in which all human beings can live to the maximum of their capacity.'

'Do you see it as a man-made purpose?'

'Oh yes. I suspect that, by contrast with some of the people who talk about animal rights, I have a species loyalty – a sense of loyalty to humanity as a whole. I think one of the strengths of the sort of animals that we are is that we can create our own future. We can, in circumstances not of our choosing, transform the world in a conscious and deliberate sort of way. I mean, we are doing that all the time. Can we actually do it in such a way as to maximize human happiness? That seems to me an important enough goal.'

I have always found it puzzling that physicists seem more inclined to be religious than biologists. Perhaps it has something to do with the fact that biologists are continually being forced to defend their ideas on evolution against non-scientifically-minded creationists. Steven being both a biologist and an atheist, I thought I would ask him whether that was his impression also.

'As you know, I am a physicist,' I said. 'My research tries to uncover the fundamental nature of space, time and matter. Are there other universes? Deep questions of that kind. It's very easy for me then to go over from the science into religious questions: What's the purpose behind it all? And I find that quite a number of my physics colleagues make that same jump themselves.

'Now I would have thought that biologists would also be concerned about ultimate questions like that, because you are dealing with *life* – which is pretty fundamental! But my impression is that biologists as a group are not really as religious as physicists. Does that tally with your experience?'

'That's absolutely right,' he agreed. 'I think that most biologists are not religious in the sense that you're referring to it. And I think there are very good reasons for that. Firstly, biology is in the throes of the most exciting scientific revolution of the twentieth century, and there's an enormous sense of optimism. Molecular biologists in particular have, I suppose, become as arrogant about the possibilities of understanding the world as physicists like Lord Kelvin were at the end of the nineteenth century. Everything was going to be explicable. So there is this arrogance about. But I also think that, as a subject, we are much more *grounded* than physics. We are dealing with things that we can see, and handle, and observe, whereas you're dealing all the time with extraordinary unobservables and phenomena beyond understanding. I think that's the reason why physicists can quite easily believe in three impossible things before breakfast, whereas biologists can't!'

'I would say, Steven, that physicists can also be quite arrogant these days,' I confessed. 'We happily talk about being on the threshold of knowing "the mind of God" and having a "Theory of Everything".'

'Yes, that is true. There is a strand within physics which says that physics is the ultimate science, and that everything else has to reduce to it. One of the things I am concerned with as a biologist is to say that physics is a model for doing *physics*; it's not a model for understanding the biological world. If you will pardon me, Russ, *we* deal with much more complicated phenomena than physicists. The brain is, I suspect, the most complex organization of matter in the Universe. That degree of complexity doesn't reduce to very simplistic physical models.'

'A popular understanding of science is that it's very confident of its findings, and that perhaps one day it will have an explanation of *everything*. Does that accord with how you see the progress of science?'

'Unlike some who think that science hasn't progressed at all, I believe that we *do* have a better understanding of the world now than we had five hundred or a thousand years ago, or even fifty years ago. But I also have no doubt that, despite the arrogance of a number of my scientific colleagues who claim that we have a true understanding of the world, there are *other* ways of viewing the world, and other ways of understanding. They come through poetry, novels, art; they come through lived experience. I think it would be misguided to ignore people's concern about religion. I am an atheist, but I am not a militant atheist in the Richard

Dawkins sense. I think it's important to try to understand why and where the religions of the past come from. I do believe they are a cry of pain in an unjust world.

'If science, as a way of understanding the world, is to move forward creatively, we have to rid ourselves of some of the suppositions which have dominated the way we have thought about the world for the last several hundred years. What's abundantly clear is that the reductionist techniques of science, as they have been developed, are unable to deal with the complexities of the social world, the complexities of the ecological interactions of the planet, or indeed the brain. We need much more comprehensive, and much less reductionist, understanding. We need much more holistic understanding. That's a new sort of science which is being forged at the moment, and I think it's the one which is going to take us forward – if we *are* to move forward.'

Moving on from atheism to a more agnostic point of view, I asked Sir Hermann Bondi, President of the British Humanist Association, what it meant to be a humanist.

'I think it is recognizing that we humans have to solve our problems and difficulties through our own resources (we cannot hope for, or expect, intervention from on high); that we are fundamentally social beings; and we must accept our weaknesses as well as our strengths.'

'Is the fact that you don't look on high for help,' I said, 'because you don't believe there is a God, or that it's the kind of God who does not intervene and help?'

'When I talk to people who have a strong religious belief, almost all of them believe in a particular revelation,' he replied. 'It is the conflicts and the contradictions between different revelations (each held to be sound by people of the highest integrity and sincerity) that makes me feel this is an area where humans are peculiarly liable to be mistaken. Of all the religions in the world, at most *one* can be right, so there must be a lot of believers who are wrong. The moment we come to any kind of religious certainty, I am exceedingly suspicious and worried.'

'Does that mean that you would still hold the door open, or at least ajar, for there being a God of some kind – although we may never understand exactly what kind of God it is?'

'I have no doctrinal refusal of the idea that the Universe may have a purpose, and may have been designed. Nobody has found a good reason for saying Yes, but equally nobody has found a good reason for saying No.'

'Is this view reasonably representative of the members of your Association, or do they take a harder line?'

'Oh, I think they cover a very wide spectrum, from people who are virtually indistinguishable from Quakers to very hard-line atheists. Humanism is, in one sense, a rule for discussion: you cannot bring an argument to a close by referring to a line in the Bible, or in the Koran, or in the writings of Karl Marx. I once had a little article in *The Times* that spoke of the arrogance of certainty.'

'If you're agnostic about the existence of God, and you've marginalized his influence (if he's there at all), does that make it difficult for humans to have an ethical code? On what kind of basis can one have an ethical code if it's not based on God?'

'First of all, I don't think anybody's ethical code *is* based on God. It is difficult; we have to rely on our own social instincts and attitudes. Of course, as a humanist I often hear this criticism, "Isn't your ethical code purely relativist?" I point out that religious people's ethical code is also relativist. Only three hundred years ago the burning of witches was considered a laudable, ethical exercise – not to speak of stamping out heresy by all means fair and foul. I would go so far as to say that what today is viewed as "Christian ethics" is in fact "humanist ethics". It selects those parts of the sayings of Jesus that fit in with our modern, humanistic attitudes. I mean, humanism is not a new invention; it has a long, and I think on the whole a very creditable history. Yes, there *are* difficulties in founding ethics. But these are difficulties we all have, and we all change our views in the course of time. When one's lived a fair time, one's noticed that attitudes have changed, and so one has to do the best one can.'

'You talked about us Christians being selective in what we

currently take from the gospels. Can you give some example?'

'Well, I have here a fairly extreme example. This is an exact quotation from the Authorized Version of the Gospel according to St John, chapter 15, verse 6. Jesus is reported to have said, "If a man abide not in me, he is cast forth as a branch and is withered, and men gather them and cast them into the fire and they are burnt."'

'It's stating very much an exclusive approach to God ...'

'Very much. And that is rarely quoted now.'

'And that's because there is today more of a realization that other religions might also be a path to the same God ... ?'

'Quite.'

'Yes, I agree.'

A mischievous look came into his eye: 'Now let me be a little aggressive! There are people around who say, "It is true that the different religions point to different revelations, and are often in contradiction, but couldn't we at least believe that which they have in common?" My attitude to this is to cite an example from everyday life. An injured cyclist is found at a crossroads, and a hit-and-run driver is obviously guilty. Four witnesses come forward. One says it was a green car speeding east, another says a red car speeding west, the third says it was a black car going north, the fourth a white car going south. We don't say, after we've heard this evidence, "At least we can be sure it was a car and not a lorry." We say that at least three of these witnesses must be untrustworthy, and we *don't* regard that as corroboration. So I would be very careful about this.'

'When I look at the world religions, the thing they most seem to have in common is the demands made by the God (or gods) on the adherent: the kind of life that person should live, based on not killing, not stealing, being truthful, honest, loving, faithful to one's partner, things of that kind. Would you say that this was a kind of basic humanist vision?'

'Yes. We must look at those tendencies as reflections of common human desires. There is a thirst for social order.'

'If one believes in God and is asked, "What's the purpose of life?", then I think the answer would be that it's to live one's life the way one believes God wants it to be lived. In the absence of God, can human life be thought of as having a purpose?'

'Yes. We are social beings and we live for, with, and through each other. And the way we make the world for our friends, neighbours, family, children and grandchildren is, I think, our purpose. It's one thing to say one believes in a God; it's quite a different thing to say one *knows* what God wants. There is a streak of certainty in that which I view with suspicion, and I think we both know of historical instances where overdoing that belief has led people into ways we regard as positively wicked.'

'So what would you say is the purpose of human life? To live one's life for the good of the human race as a whole?'

'Yes, yes, yes! ' he declared enthusiastically. 'I quote Tom Paine, "My religion is to do good, and my country is the world." He was a remarkable character, Tom Paine.'

'You are against religious people being certain about their understanding of God. How would you react to scientists who speak as though science was their God, and that they were on the road to a complete understanding of everything, total certainty?'

'You will not be surprised when I say I laugh – if it doesn't raise my blood-pressure and make me furious!'

'And do you think this will always be the case, no matter how far one projects into the future?'

'Science advances because we get new and better experimental equipment. So the advance of technology governs the advance of science. People who say science is primary and technology is secondary don't know what they are talking about.'

'In my own research field of high-energy nuclear physics,' I ventured, 'I am very dependent on the fact that I must have big machines. The present one is 27 kilometres in circumference. It does occur to me that, in order to put the last jigsaw piece together, it might actually require a machine which is bigger than the solar system. There is no reason why the uncovering of scientific laws has to be geared to our Gross National Product.'

He laughed. 'Very nicely put.'

'Certainly, my experience is that as you build these bigger

machines, they answer the questions that you were asking, but they always throw up new questions that you hadn't previously thought of.'

'I think we are very much at one on this,' he concluded.

Science and religion together

So far we have heard from people who, if not positively antagonistic to religion, nevertheless consider it as being largely irrelevant and as having nothing to do with science. How about more positive views of the relationship between the two?

Biblical scholar Ernest Lucas had earlier pointed out to me that the Genesis account of creation was in terms of an ordered cosmos; it was possible to make sense of it. This he had contrasted with other primitive creation accounts, like the Babylonian one which was based on chaos. One of the prerequisites for engaging in science is the assumption that there is order and intelligibility in nature. I asked Ernest whether he thought the Bible had played a part in bringing about a frame of mind that would later prove fertile ground for the rise of science in the West.

'I think it did,' he affirmed. 'Historians of science have increasingly recognized this. The early modern scientists, such as Kepler, Bacon and Newton, had confidence that they could go out and study the world and find order – and what's more, an order they could understand. For them, that confidence came from their Christian beliefs. Secondly, they believed they were made in the image of God, and therefore their minds would be able "to think God's thoughts after Him", to quote Kepler's famous words, and find that order. So there were two factors there: one being the belief that the world is ordered, and the other that I can understand it because I have something of the mind of the Creator in me.'

Following up this line of thought, I asked Sir John Houghton, 'If one is to regard the Universe as in some way reflecting the

Intelligence that put it in place, can we hope to learn anything about the nature of God by studying his world?'

He responded, 'The old reformers and the early scientists talked about the "two books": the book of nature, and the book of the Bible which tells us about God's revelation in Jesus. It's very important to look at both of them. I sometimes think it's like having binocular vision. You look through one half of the binocular and it's not very good; you have to look through both to get the real depth of picture.'

Although the world is basically intelligible, it does incorporate an element of 'chaos'. I raised this with Sir John. 'During your term as head of the Met Office, you must have taken an awful lot of stick from time to time when you got your weather forecast wrong. We now know that in some situations, like the establishment of weather patterns, very small local changes can be subject to big multiplication factors and lead to big consequences. This is what we now call "chaos theory", or the way we are incapable of predicting the future accurately, even in principle. Can you explain how chaos theory affected your work?'

'Yes. We could start with the storm of 16 October 1987, which is seared on my chest!' he smiled ruefully. 'I remember it very well; we didn't do a very good forecast. It was a very unusual storm. Later, we studied it a great deal. It was interesting to find out that actually we *could* have forecast it quite well, had we had use of *all* the data that was in principle available to us, and had we had some rather better techniques in the computer model. But that doesn't mean, of course, that we can in principle forecast all weather. After one or two weeks, chaos takes over ...'

'... the imprecision of one's initial measurements gets multiplied up?' I asked.

'That's right – exponentially. So you can't go beyond about two weeks in ordinary forecasting at these latitudes. The climate system is not entirely chaotic; it does respond in a regular and predictable way on average, in a statistical sense. We say the weather and the climate system is *partly* chaotic. And a great deal of what occurs in the natural world is of that kind. There are some things we can't predict, but there are others which come out as really quite orderly and predictable.'

'Would you say that chaos theory is an invitation to us to be more humble when it comes to science and what science is able to do?'

'In a way, yes. It demonstrates how complex the whole thing is. There are horizons beyond which we cannot see. There is enormous flexibility and variety and diversity in nature, on a scale we just can't imagine. That's a message which is a bit different from what Newton gave three hundred years ago when he was talking about planets going round the Sun. It is a very important message for people to get about science – and perhaps about its connections with other sorts of knowledge.'

'Does that tell you anything about God?'

'It perhaps gives us some idea of the size of God and the scale on which he operates: the diversity and the enormous variety he's put into not just living systems, but also the Universe as a whole. If we can stand back from our science and look at just how big and how fascinating and how complex it is, but also how *orderly* it is, we do get some picture of what God must be like.'

The idea that one can look to science to throw light on God has been advocated by physicist Paul Davies. I reminded him that he once wrote that he saw science as a surer path to God than religion.

'What did you mean by that?' I asked.

'First of all, I was being deliberately mischievous!' he said with a grin. 'I felt that we should provoke a debate. Let me put it this way. Science and religion, at least as they're being practised in the major institutions, come at the subject matter from opposite directions. Religion is usually based on some sort of doctrine or an ancient text which is meant to contain revealed truths. It's something you have to accept on faith. And you're not supposed to change these truths – although in practice we all know that it has happened, that the Bible has been endlessly reinterpreted to fit the scientific facts. Science, on the other hand, starts from exactly the opposite point of view. You have to accept, as an act of faith, that there's an existing order in nature that is intelligible to us. That's a huge act of faith. But once you've done that, everything else is tentative or provisional. At any given time, even though we might feel confident we have a good understanding of the Universe, we must always as scientists be prepared to change our minds in the light of new evidence. So whilst individual scientists are certainly dogmatic, and certainly have passionate beliefs that they adhere to, the scientific community as a whole sets uncompromising standards of rigour and discipline. When the evidence suggests that a new concept is called for, or an old theory must be rejected, then we simply have to swallow hard and do it. So we have one system of thought which is based on rather rigid origins, and the other which is open, exploratory and tentative.

'Now a lot of people think that because scientists are always changing their minds, for the reasons I've just explained, you can't really rely on science. They'd much rather put their faith in an ancient book. I think it's exactly the other way round. The power of science is precisely because it's adaptable; it improves all the time. Scientific enquiry mirrors, albeit in some imperfect but improving way, a really-existing order in nature. I think evidence for something like meaning or purpose or design in the Universe comes from the demystification of nature through scientific enquiry. Science leads us in the direction of *reliable* knowledge. That's the essential point I want to make. In spite of the fact that it's tentative and provisional, generally speaking it is reliable knowledge – and that's why it's a more reliable path.'

'I wonder, though, if you're being fair in drawing such a clear distinction between scientific method and religious method,' I objected. 'Certainly there are many religious people whose reaction is very much as you describe. There is the Book, and it must not be changed ...'

'That's the word of God ...'

'... Yes, that's the word of God,' I continued. 'But if you take the Old Testament and you put the books in chronological order (as we best understand that to be), you can see that the idea of God is actually developing, and improving, as time goes on. He starts off as being just one God amongst many gods. He's only interested in the Israelites; he doesn't care a fig about the Egyptians (he has them killed off). Then he becomes just *one* God. He starts off by being a God who is very vengeful and wrathful, and becomes ...'

'He mellows with time!'

'He mellows with time. He eventually ends up as a *loving* God. Now it seems to me that here you have a kind of development and reinterpretation as you're going along, with a more and more adequate understanding of God, which is rather similar to how you describe science progressing.'

'I agree entirely. And it seems to me that in my discussions with theologians, those who are prepared to regard the Bible as what it is – an interesting collection of ideas put together over a period of time (much of it when people had only a rudimentary understanding of the natural world) – and to accept that it's got important insights for human beings but is not in any sense *God's* description of the world, then that's fine. If you regard theology as an exercise in the evolution of ideas, then I think we do have something very similar to the scientific process.

'But of course, many "ordinary" believers don't see it that way. It's often said to me, "Isn't there a huge gulf between science and theology?" And I say, "No, behind closed doors scientists and theologians get on very well indeed." We all agree that the Universe began with something like a Big Bang, and that life has developed through evolutionary processes, and so on. There isn't really any quibble over the facts of the world, though there may be disagreements about the *interpretation* of those facts.

'The real gulf is between theologians and ordinary believers who still cling to the Sunday-school notion of God as a cosmic magician who works miracles from time to time, just like any other force or agency in the Universe – a very uninspiring view of the Deity, in my opinion. I think it's up to the theologians to come out of the closet and take their message to the people.'

I next turned to John Habgood. I asked him, 'When you look at the advance of science, is there anything there which radically changes the kind of God we can believe in these days?'

'Not radically, I don't think,' he replied. 'It's interesting that whereas nobody goes back and reads old scientific texts (not unless they're historians of science), we all constantly go back to the Bible and read it. There we find a living presence of God which is just as real as it was in any other age. So there are enduring insights into the nature of God which don't change.

'But some things *do* change. One which was brought to the fore by the theory of evolution was a greater sense of God's continuous activity in relation to the world, the so-called "doctrine of continuous creation".

'Also,' he added, 'science has made a good deal less plausible the notion of a God who is, as it were, continually interfering in the course of human life. To some extent, our modern understanding of the world does emphasize the *distance* of God from his creation, rather than that very intimate sense of God's presence everywhere which seemed more prevalent in New Testament times, and right through into medieval times.

'Theology always has to take into account the culture within which it is trying to think seriously about the nature of reality. To that extent, theologians and Christians generally need to be listening to what science is saying to us about its picture of the world. But they should not necessarily be taking that as the *final* gospel for the way things really are. There are other depths which it's the task of theologians to explore.'

'Do you see any areas of scientific endeavour where we ought to slow down and be careful – for example, certain aspects of genetic engineering?'

'Yes, I do. There have to be limits on what can be done scientifically in terms of method. In terms of knowledge, it's very difficult to set limits. Certain types of knowledge are very hard to handle. That is why it's so important that, alongside the growth of science, a massive effort needs to be put into trying to work out the ethical and social implications of what's being discovered, particularly in genetics.'

*

I put it to philosopher Roger Trigg that one of the perennial problems with religion is arguing about whether it refers to any objective reality. Is there actually a God out there, a God who would exist whether or not there were people behaving religiously? In science, it seems to be somewhat different. Scientists are generally thought to be trying to describe a world which really *is* out there. How did he see objective reality in both the religious and the scientific spheres?

He began cautiously. 'First of all, I think that it isn't absolutely agreed – surprising as that may sound – that scientists are discovering one objective reality. Working scientists assume that; they are "realists". In other words, they assume there is an independent reality. But a lot of theorists of science – sociologists and philosophers – have more recently been emphasizing that scientists are in the business of constructing theories, constructing models. Science is a social practice, like any other social practice; scientists form a community. They build a picture of the world. Indeed, they "construct" a world, and people sometimes talk about the different worlds different scientists may live in.

'This perhaps isn't a problem that religion alone has. It's a very basic philosophical issue: are we talking about reality, or are we building up pictures of reality that are in a sense reflections of the human mind? It's like looking down into a pool and seeing our face looking up. Is the picture we build of the world a reflection of us, or are we looking at something independent? I personally think that scientists are in the business of discovering the world, not constructing it.

'Now many people in the field of religion nowadays believe very strongly that religion is constructing rather than discovering. But I think it's an absolute presupposition of religion that it is talking about an independent God. Either there is a God, or there isn't. Either the physical world has a particular nature, or it hasn't.

'I think science and religion are each dealing with the same reality. We all live in one world. Science is trying to discover it, but it is a world which, if religion is right, was created by God. Therefore I would expect science and religion (although they have different agendas) nevertheless to be saying things that may be relevant to each other. I'm always very unhappy about the kind of religious belief that fences itself around so that nothing science ever said could disprove it, be a pointer against it – indeed, even be relevant to it.'

'So are you saying that science and religion are both trying to discover a reality out there, but that the ideas we come up with are bound to be coloured by social pressures, our cultural and historical background – and that this applies to both science and to religion?'

'Yes, that's true. We are influenced by our background, by the concepts we have, the way we think.'

'If both science and religion are culturally influenced, then they're not giving a crystal-clear picture of the objective reality they're relating to,' I agreed. 'When one looks at religion, one can see that happening very clearly: there are several world religions. Although there are common features in what they're saying about the transcendent out there, there are obvious differences too. In science one can see *some* differences: for example, with medical science in China you've got a system incorporating acupuncture, a practice that has no place in Western medicine. But the cultural differences don't seem to be anything like as marked as they are in religion. Is there any particular reason why these cultural differences should be so marked in religion?'

'If one starts with the idea that there is one reality, and then people disagree about what it's like, it suggests that all or some of us are wrong. The fact that there are different religions, each making its different claim, should make us rather humble. It doesn't follow that I'm right and they're all wrong, or that they're right and I'm wrong. But we should be willing to look at the truth together. When you say that in China they practise acupuncture, I don't think, "Well, that's just a funny thing the Chinese do". If there's something to be learned there by Western medicine, we ought to learn it. I suspect there might well be something to be learned.'

'If we see religion as trying to discover the nature of God, and science as trying to discover the nature of the physical world, do you see any connection between those two objective realities out there?'

'When you take part in scientific research, you're looking at a world that you're assuming is ordered. You're assuming there are regularities to be seen. Now why is that? In a way, it's remarkable that the physical world should be so ordered. Is it just a matter of chance? Certainly, we couldn't do science if we weren't able to discover order. I would have thought that the order science discovers in the world in some way reflects the mind of a creator behind things. In other words, the order has a religious base. God has somehow created a world which shows something of his own mind, his own rationality.

'One of the most interesting things about mathematics is: why is it that when people are just sitting in a chair writing things down on a piece of paper, or getting up and writing them on a blackboard, that this somehow says something about the workings of the physical world? How is it that the thoughts of mathematicians somehow reflect the innermost mysteries of things physical? Mathematics is absolutely vital for physics, yet why should the physical world reflect that kind of order which can be put into mathematical form?

'One answer would be that we, as creatures made in the image of God, reflect in a very small way the rationality of the God who made the world, and that the rationality we possess somehow reflects the reason built into the very scheme of things.'

'If you see rationality as the connection between God and the physical world, does that mean that we ought to look upon theology as a branch of science? Should theologians investigate and experiment on God, in the same way as scientists go about their work in the laboratory?'

'I think that theology – which, after all, was once called the "Queen of the Sciences" – has one thing in common with science: its subject matter is governed by the nature of objective reality. But it isn't right to think that theology is just another physical science, and that we do it through ordinary experiment – so that if we can't find God in the laboratory, he isn't there. That's the mistake that the philosophers called "logical positivists" made. They tended to think that if you couldn't verify something, it wasn't meaningful.

That's making science the arbiter of what's there. Now it certainly *is* the arbiter of what's there physically. But if you believe, as I do, that there may be other realms of reality, then science isn't going to answer the basic question about what is there. There may be other ways of discovering other kinds of reality that may still be real and still be important.'

'Is one of the problems the fact that one believes in a God who is personal, and we know what it's like to be a person? We can make our own decisions; we can decide how much of ourselves we're going to reveal to other people. Do you think that one of the problems a theologian faces is that he's dealing with a God who is personal, who can decide what he's going to reveal of himself – whereas the scientist is dealing with an impersonal reality which has no alternative but to yield up its secrets if the scientist asks the right questions and does the right experiments?'

'I certainly think that the theologian is very much subject to the possibility of revelation,' he agreed. 'If God never chose to reveal himself to us, we would never know anything about him. But on the other hand, I think that God also reveals himself in the nature of the physical world. If the physical world is the handiwork of the creator, in a sense we're learning something about God even as scientists working on the physical world. Perhaps it's a mistake just to see the physical world as something inert, totally separate from God. In one sense it's separate, in that God created it and doesn't depend on it. But it may somehow reflect something of the nature of God. Indeed, there is a whole area of theology called "natural theology" that tries to reason from the nature of the world to God. So I don't see science and religion as totally separate. Science itself may be a path to religion.'

This seemed to me to be very much in the spirit of what Paul Davies was saying.

A theologian who has a particularly strong interest in science is Nancey Murphy. I asked her how this interest arose, and in what ways she saw science as enriching her understanding of theology.

'Part of the explanation,' she told me, 'is that I was going to school during the Sputnik era. Americans thought it was terribly important that we should all get good at science, so that we'd be terribly good at building weapons. Also, I had an older sister who was a science fan, and I was trying to keep up with her. But these days, having studied some science – although I'm certainly no expert in the field – I can't imagine what my life would be like if I didn't have knowledge of the world around me and how it works. It would be almost comparable to being illiterate and living in a city full of street-signs, and not being able to read them. So I treasure the knowledge that I've got of the natural world, and I've come to think that it's part of God's will that we should have this knowledge.

'God created us in God's own image and likeness. There's a lot of disagreement about what that means. At least part of it is that we have minds capable of knowing God. But also of knowing what God has *done*, appreciating the creation that God has produced, the complexity, the beauty and the order of it. So I think it's part of our mandate as creatures of God to learn as much about the natural world as we can.'

'When you as a theologian look at the totality of science, what do you think it is saying about you and me – the dignity of humankind?'

'It's allowed us, or perhaps forced us, to recapture the notion that we are part of the natural world. That's certainly a Biblical teaching – the notion that we're made out of the dust of the Earth – but for a long time the human race tried to deny that. But science has forced us to recognize that fact again, and that's good.

'The complexity of the human person is marvellous; we have learned to appreciate the cleverness of God in creating beings like ourselves who are so wonderfully complex.

'Some people say that the vastness of the Universe shows our insignificance. But if you are assuming, as I do, that God created the Universe in order to have creatures like ourselves who could know him and respond to him and love him, then the way to look at it is, "Look at the huge Universe he had to create in order to have situations suitable for us to emerge!" So the very vastness of the Universe can be turned around to reflect on the importance of ourselves in God's eyes. If of course you are assuming, as I do, that God really wanted us to be here.'

Science, religion, and education

From what we have heard, it is clearly the view of many working in the field that science and religion *can* be reconciled. That being so, why does the idea of a conflict between the two persist among the public at large?

I suspect it is partly due to a general lack of appreciation of what really matters in religion; what one is actually supposed to be getting out of the Bible, for instance. As we have seen, it is all too easy to get hung up on a literal interpretation of writings, some of which were probably never intended to be read in that light in the first place. Here we have in mind not only the creation stories, but also possibly some of the accounts of miracles. Taking religion seriously is *not* a matter of accepting, as a package deal, one particular way of interpreting Scripture and the tradition of the Church. But that is the misapprehension under which many people labour. Out of a sense of intellectual integrity, they cannot bring themselves to believe what they *think* they are being asked to believe, and in the process they deny themselves access to religion's timeless truths.

But that is only part of the story. It can also be difficult to get a proper perspective on science, on what it can and cannot say. To some extent, we scientists must plead guilty to adding to this confusion. Take, for example, the way physicists have recently begun to speak of the prospect of one day achieving a 'Theory of Everything'. What is the public supposed to make of *that* when they read about it in their newspapers? On the surface, it appears to be a claim that all knowledge and understanding is about to be swallowed up by physics; there will then be no need of religious, or any other kind of discourse. But that, of course, depends on what one means by the word 'everything'.

Fellow physicist Chris Isham: 'I think you've put your finger on the crucial point,' he told me. 'You have to limit your notion of "everything" before you can really start to talk about these things. What physicists really mean is that maybe the time will come when, as far as the notion of microscopic matter is concerned, we have a single theory which seems to describe everything that goes on. That may well be possible; I see no reason why it shouldn't. In that *limited* sense, I think it may be feasible to have a theory of everything. But of course, if one means "everything" in the sense of the totality of all human experience –wisdom, understanding, perception, and so on – then I think it's a non-starter.'

'Even if we did get the Theory of Everything – in the limited sense in which you are describing it – would we be able to be sure that it *was* the ultimate theory?'

'Clearly not, because at the end of the day you would have to test it. You cannot test everything. All you would know is, as has always been the case with theories of physics, that it works within the domain where you've tested it. And that's all you can *ever* deduce.'

A sign of the increasingly positive attitude towards science and religion to be found in some quarters has been the recent setting-up of a lectureship in the subject at Cambridge University. I asked Fraser Watts, its first holder, how the post came about, and how he saw its significance.

'It arose out of the wish of a number of people here in the university to have a post on the relationship between science and religion, and out of the generosity of Susan Howatch in making the money available to do this.'

'She's the novelist?'

'That's right. And from a personal point of view, I was just delighted to have the opportunity of working on the background of my theological and scientific interests.'

'And how do you see your role here? Do you have any special mission?'

'It is enormously important to be looking at theological questions in ways which are appropriate to our scientific age. But the particular contribution I think I have to make to that is my back-ground in the *human* sciences. A lot of the running in theology and science in recent years has been made by people with a background in physics. And clearly there are some very important questions to be addressed there. But the questions in the human sciences have to some extent been neglected, and I see it as my particular mission, if you like, to address those.'

'In your capacity as a lecturer in science and theology, what do you see as the public perception of the relationship between the two? Is it a healthy perception, or a mistaken one, or what?'

'There is a widespread perception that science and religion are in conflict; that science has disproved religion; that religious faith is intellectually disreputable. This perception comes from the latter part of the nineteenth century. It doesn't date back any further than that, but it is the received wisdom. Now I think that's an entirely unjustifiable set of assumptions. If people examine things more carefully and get a sophisticated view of what science is, and a good understanding of what religious faith is, then the idea that they are in conflict just drops away.'

'So are you saying that it's a misunderstanding of both the science and the theology that's to blame? It's not just one?'

'That's right, it's both. It comes out of a rather simple-minded and grandiose view of what science is – that it is the *only* way of studying anything. But the philosophy of science has got a lot more subtle over the last twenty-five years. And though science is still, in my view, to be seen as enormously important, it is *not* the only way of talking sense about things. We need a more modest view of science than some people have had in the past. We also need a more subtle view of theology and the interpretation of the Bible.'

'Do you find the situation any better at the academic level? Are theologians in good conversation with scientists, thrashing out the implications of scientific discoveries for theology?'

'Yes, things are a little better there, but not quite as good as I would like them to be. Certainly I find in theologians a lot of enthusiasm for the idea of having a dialogue with science, as indeed with a variety of other disciplines. There's been a most fruitful interchange at the level of examining how things are known in science, how things are known in religion, and what the source of our authority is.'

Though there are currently many healthy signs of a dialogue developing between professional scientists, theologians and philosophers, it is difficult to perceive much constructive discussion going on at the level of the general public. This is perhaps not surprising. Without access to the right kind of information, it is difficult, if not impossible, to arrive at a balanced judgement concerning, for instance, the status of scientific pronouncements.

John Durant has a special responsibility in this area. He is Professor of the Public Understanding of Science at Imperial College, and an assistant director of the Science Museum in London. I met up with him at the museum, and put my concerns to him.

'When one hears about scientific discoveries such as the theory of evolution, modern ideas of cosmology, the vastness of space and that sort of thing, one cannot help reflecting on what all this implies for the status of human beings; where we fit into the scheme of things. Do you think that scientists have a duty not only to explain the science itself, but also to help people to understand what the *implications* of science are for how we should view ourselves?'

'Yes, I think that's fair,' he replied. 'It is difficult not to agree with the idea that scientists ought to go beyond the mere reporting of new information and say something about what it all means. But I think there's a trap there as well: scientists are not always well placed to assess the meaning and the implications of what they've done.

'One of the functions of metaphysics, as it used to be called – it has become an unfashionable area – was precisely to consider the deeper significance of what we know, or *think* we know, about the world. One of the problems we have today is that the technical success of science has led at least some scientists to be a bit over-confident of their ability to judge the wider significance of what they've done. Amongst scientists, I don't see any sophistication in metaphysics that could be said to match their technical sophistication.'

'So does this call for another kind of person: someone who will concentrate on the implications of science? Perhaps a former scientist who is no longer spending all his time doing research, but is getting more familiar with philosophical and religious ideas?'

'Yes, I agree with that in part. Frankly, there is a continuing need for philosophers. Philosophers are there, it seems to me, to encourage us to ask the right sorts of questions in the right sorts of ways, and help us think about how to find answers to them.

'There's also a case for trying to educate scientists a bit more broadly than we do. I don't think it's helpful if very technically accomplished scientists arrive at mid-career, at a time when they're making some of their most important contributions to science, and are completely innocent of the larger world of ideas and concerns with which their work interfaces. There's a strong argument here for the kind of broad-based education that some other countries, like France and the United States, offer.'

'Am I right in saying that you are unique in being a Professor of Public Understanding of Science?'

'Yes, I get rather worried when I'm asked that,' he responded with a smile. 'As far as I know, I am the only one of my kind. I feel like a potentially extinct species! I don't know of any other person who has quite that title. But there are people with appointments in "science communication" – which is not a million miles away.' *

'Do you see it as part of your remit to get the public to think about the *implications* of science, or do you just see it as helping the public to understand the science itself?'

'I would take the broader view. We have good reason to think that many people are interested in the wider implications of science. I mean, why do books on cosmology sell so well? It can hardly

* Since this conversation took place, Richard Dawkins has been appointed to a newly established professorship of the Public Understanding of Science at Oxford University.

be because cosmology has great practical relevance to everyday life. It seems to be, to use a grandiose phrase, the human quest for meaning. We wish to understand ourselves and our place in the Universe. I think we should respect that interest. It's a sensible interest for citizens to have, and I do think it's part of the responsibility of the scientific community, and of people like me, to try to facilitate a more informed and a more constructive discussion of those subjects.'

'There are occasions when scientists do leave their laboratories and speak about the wider significance of their work – particularly for religion – sometimes making claims that it backs up religious beliefs, and on other occasions that it refutes them. What do you think about these pronouncements?'

'They are interestingly diverse. Any lay person looking in from outside would immediately be struck by the fact that scientists do not by any means speak with one voice on this subject. You have a number of scientists, particularly from the world of the physical sciences, who speak, apparently quite confidently, of the harmony between science and religion, indeed of the sense in which science is providing new evidences and illustrations of religious truths. And you have equal numbers of scientists, not all of them physicists, who speak equally confidently about how science has made religion redundant, if not absurd. I suspect scientists are as various in their views of religion as the rest of the community.

'Historically speaking, there isn't very good evidence for a neat relationship. There is evidence for a very close set of relationships, but they are complex. I think that most simplistic formulae are likely to be out of order. I confess to some disquiet when I see what are almost tub-thumping approaches to this by some scientists. One or two scientists seem to have become professional religion-bashers. They seem to be attacking religion as it were almost *on behalf* of science, or with the *credentials* of science. I do get uncomfortable about that, and a lot of other scientists do as well. I just don't think that science is in the business of proving or disproving metaphysical and theological propositions.'

'As I go round the Science Museum, I see a lot of very good

explanations of the sciences. I don't see much, if any, about the wider significance of these developments. Do you think a museum like this has a duty to go beyond simply explaining the science?'

'Yes, very much so. Speaking as it were on behalf of the Science Museum, I would want at first to plead partially guilty. I think in the past, museums like ours have tended to concentrate on technology rather than science (that, I think, is something we *don't* need to apologize for – technology lends itself to exhibition, and it's important). But the museum has also tended to concentrate on the clear-cut, the unproblematic, as it were. In the past, visitors to a museum like this might have got the impression that science and technology are an accumulation of straightforward and universally accepted truths and practical applications. The difficult questions – moral, social, political, economic, religious – have not traditionally been part of the agenda of most science museums.

'They are now getting on to our agenda slowly but surely. Certainly we now take a much greater interest here in the issues of science, including moral and religious ones, than we did in the past. But because of the pace at which museums can afford to change, you still don't see a lot of it.

'An example perhaps of the new emphasis is our most recent gallery, "Health Matters". It's about twentieth-century medical science. I don't want to overclaim for what we do there. Much of the gallery is rightly concerned with the impact of new science and of new techniques in medical practice. But there is a whole section devoted to three major areas of disease – heart disease, cancer and Aids – and that section actually includes a lot of the responses that patients, and groups concerned for patients' interests, have had. These are not just technical responses or practical responses, but *emotional* responses to those conditions.'

'Science needs to attract public funding in order to finance its research, and incredibly large sums of money are required. So is there not a tendency for science to oversell itself – to be overly optimistic and triumphalistic – simply in order to get these funds? Might it not be downplaying the *problems* that might come from scientific investigation and its limitations?'

'That's always a risk. But I suppose the political process is reasonably accustomed to special-interest groups in society banging the drum very loudly on their own behalf. I guess the world of the arts tends to play up the cultural and national significance of the arts when it's making bids for money, just as the world of science does. So I'm reasonably confident of the political process being able to read between the lines and compensate for science's understandable salesmanship.

'Perhaps when it comes to *public* pronouncements, though, we need to think more carefully. Science still has – quite rightly, perhaps – a very high reputation amongst the public for credibility, trustworthiness and reliability. We know from studies of public perceptions that scientists still rank relatively high amongst different professional groups in terms of public confidence – unlike politicians or journalists, for example, who rank notoriously low. But I think that if there's too much over-selling going on in the public domain, that could perhaps lead to some long-term undermining of confidence.

'A good example would be the nuclear industry, which in the immediate post-war period was presented to the public as an extraordinarily problem-free and positive contribution to society: "atoms for peace", "energy too cheap to meter", and so on. And as we know, things didn't work out quite like that. The scientific community in general ought to be learning the lesson that there may be a price to pay for overselling yourself.'

'We know that in the past religion has had a great deal to answer for. Do you think that science has things to apologize for?'

'Yes, to be frank, it does. And it would be a miracle, I think, if it didn't, in the sense that science in recent times has become very powerful. And to the extent that scientific knowledge is powerful, it finds itself necessarily enmeshed with all those other areas of human action which are morally complex and morally ambiguous. We live in an age when the applications of science are morally ambiguous. We can all think, I guess, of examples in the twentieth century of science used for great human benefit. Penicillin would be my first example. Equally, science has been used for great

194

human *dis*benefit. The atomic bomb would be my first choice there. And they both happened at roughly the same time. The scientific community now generally knows, and the public generally knows, that we have to live with the responsibility for the use of our scientific accomplishment.

'I do think, though, that there is an underlying optimism. If you were ultimately pessimistic about the ability of human beings to deal morally with the world, I can't see why you would remain committed to the scientific process. There is a certain sort of optimism behind science which says yes to all the moral ambiguities and complexities and difficulties. It's better to know than not to know. We must hope and pray that we shall have the moral wisdom to make good use of this knowledge, and not to abuse it. Most scientists I know have that underlying moral optimism.'

Michael Poole has made a study of science education. Meeting him at King's College London, I asked him how he saw science in relation to religion.

'There are four positions often held,' he said. 'One of conflict, one of uneasy truce, one of peaceful coexistence, and one of dynamic interchange. I like the last one. I think the discoveries of science can give useful ideas to theologians when they're discussing, for instance, how the Bible is intended to be read.'

'In recent years there have been quite a number of books published by scientists (particularly cosmologists) who, having described their science, tend to wax philosophical in their last chapter. What do you think about the competence of the average "brilliant scientist" to reflect on the significance of what they've been describing?'

'The Last Chapter phenomenon is well known!' he laughed. 'Again, one doesn't want to be particularly critical here, but there is a tendency in those closing paragraphs for authors to get carried away into areas they're perhaps not so familiar with – trying to draw conclusions from the science that really are neither philosophically or theologically tenable.

'What is interesting about all this popular literature is that it is at least raising awareness, and showing that these questions need to be asked. That's useful. But sometimes the kind of mistakes being made do seem rather naïve. For instance, it's quite common to find the claim that if we have a complete scientific explanation of something, what need do we have of God? Which of course is nonsense.'

'What do you reckon is being done in schools to try and prevent a false understanding of the relationship between science and religion?' I asked.

'Not as much as I'd like to see being done. But there are initiatives being taken: for instance, the agreed syllabuses in religious education (there are units looking at the interplay between science and religion). In science education, the Education Reform Act specifies that one should be looking at the spiritual, moral, cultural and social aspects. And the Office for Standards in Education, in its inspection schedule, makes the point that it is a *whole* school curriculum matter to promote the spiritual, moral, cultural and social. So that means that science has its part to play as well.'

'How do we get science teachers to the position where they can actually fulfil that obligation? Presumably a science teacher is not necessarily trained in the *philosophy* of science, in how science fits into the whole scheme?'

'No, this is the problem. Unless a first degree in science has had a component of History and Philosophy of Science – and some universities do have this – then people come along who've got a large *content* knowledge, but perhaps not so much background knowledge in the history and the philosophy. Only things like in-service courses can help teachers in this.

'Largely speaking, it's a matter of good science education. If one is educating well in science, one shouldn't be conveying ideas like "scientific explanations are the *only* explanations" – which is one of the things that has bugged the science and religion debate for a long time.'

'So are you advocating that science lessons shouldn't just be a case of putting across the information about the science; the teacher's really got to be able to place it in the context of our wider understanding of other aspects of life?'

'That's right. It's a legal requirement.'

'How about religious education lessons? Are RE teachers equipped for explaining how science and religion are to be related to each other?'

'Well, this is the other side of the coin. The RE specialist won't have done very much science or philosophy of science, though they will probably have a better background in philosophical matters than the science specialists. So there's a place for training RE specialists in the science area, as well as science specialists in the philosophical areas.'

'Basically, you're criticizing the way we've been taught in the past. Science people are taught science, RE people are taught RE, and never the twain shall meet. Really, you're looking for a much broader kind of education, one where people are able to cross boundaries. Is that what you're after?'

'Yes. One's looking for an integration. I don't want to be too critical, however, because it's not easy to cross boundaries. Interdisciplinary studies are always difficult. I remember, for instance, when I was teaching in a south London comprehensive school many years ago. I was in the physics lab. There was a door to the adjoining chemistry lab. My colleague, the chemist, and I used to laugh at how sixth-formers, as they crossed that dividing door between the two labs, changed their whole attitude towards atoms and molecules. Although one did something like electrolysis in physics and in chemistry, their whole perception of the process seemed to be dependent upon which lab they were in. If one has that kind of a split between two so closely aligned subjects as physics and chemistry, then what about the feeling of separation there is in the minds of pupils between the sciences and the RE department at the other end of the school!'

Conclusion

That concluded my series of conversations. How to sum up? What, in a nutshell, is the present state of play in the science and religion dialogue? My overriding impression is that we have to make a clear-cut distinction.

At the *academic* level, there is much good and constructive discussion going on. There is no consensus: as you will have gathered for yourself, there is every shade of opinion among the professionals as to how one might view the relationship between science and religion. But at least the debate is well-informed, and becoming increasingly so. This is evidenced, for example, by the establishment of the lectureship held currently by Fraser Watts; the founding of the Center for Theology and the Natural Sciences by Bob Russell in California; the setting-up of the Institute for Science and Religion, directed by Phil Hefner, at Chicago; the recent award of the John Templeton Prize for Progress in Religion (worth $1 million) to Paul Davies for his contributions to science and religion; and the listing in *Who's Who in Theology and Science* of over 1000 professionals actively working and publishing books and papers in the field. There are many other indications one could cite to support the claim that the investigation of the science–religion interface is one of the fastest-growing areas of academic study today.

On the other hand, when it comes to engaging the wider public in the exploration of these ideas, there is much to be desired. From their late teenage onwards, the majority of people seem all too willing to settle for an inadequate perception of the issues, founded on misconceptions of what both religion and science are saying, and *can* say. This assessment then becomes the basis for attitudes carried into adult life.

There are, as we have seen, a few encouraging signs that through the reform of school syllabuses, the suitable training of teachers and the establishment of lectureships on the public understanding of science, we might one day move beyond the present situation. But it will be an uphill struggle. An essential ingredient for appreciating both science and religion is the sense of *wonder*. Yet in my work of bringing the topics of science and of religion to a wider audience, I have been struck by the marked change that comes over young people as they enter their early teens. Too often, an air of bored indifference sets in and this precious sense of wonder is to a large extent lost, never to be regained.

I use the word 'wonder' in the first place to refer to that childlike emotional reaction to something awesome and mysterious – a response that can lead to worship. Secondly, it refers to the way we are led to enquire, to wonder, what it all means. In this connection, one is reminded of the child's endless stream of questions beginning 'Why?' Wonder, in the sense of curiosity, is when we move beyond pure emotion to where the intellect is engaged. Both types of wonder become submerged in the early teens.

My own response to this has been to spend an increasing proportion of my time writing for, and speaking to, audiences of young people of twelve years old and less. I find it easier to establish a rapport with them than with many an audience made up of the average late-teenager or adult. The sense of wonder, essential for my scientific work and my religious life, is something these young people have in natural abundance.

The loss of wonder in later life is *not* inevitable. It does not have to be this way. As amply demonstrated in these conversations, the scientists I have spoken to on my journey still possess it, in the sense of being emotionally affected by the power and mystery of the world they are discovering. Some, indeed, go further and show a continuing willingness to go beyond the 'how' questions in pursuit of the 'why'.

Jesus said that one had to become as a child in order to enter the kingdom of God. The same attitude of eagerness coupled with humility also appears essential in science. The re-establishment of an honest, genuine openness to new challenges and ways of thinking is perhaps the first and most important step towards a better understanding of science, of religion, and of the way the two relate to each other.

Index

Page numbers in **bold** refer to conversations with the contributors to this book.